56th Annual Conference of Microscopy Society of Southern Africa

Proceedings of the Microscopy Society of Southern Africa
Volume 48, 2019

Langebaan, South Africa
1 – 5 December 2019

ISBN: 978-1-7138-3561-5

Printed from e-media with permission by:

Curran Associates, Inc.
57 Morehouse Lane
Red Hook, NY 12571

Some format issues inherent in the e-media version may also appear in this print version.

Printed with permission by Curran Associates, Inc. (2021)

For permission requests, please contact Microscopy Society of Southern Africa
at the address below.

Microscopy Society of Southern Africa
1 Meiring Naude Rd, Brummeria Pretoria
0001, South Africa

Phone: +27 12 841 3643

www.microscopy.co.za

Additional copies of this publication are available from:

Curran Associates, Inc.
57 Morehouse Lane
Red Hook, NY 12571 USA
Phone: 845-758-0400
Fax: 845-758-2633
Email: curran@proceedings.com
Web: www.proceedings.com

MICROSCOPY SOCIETY OF SOUTHERN AFRICA

PROCEEDINGS
VOLUME 48, 2019

CLUB MYKONOS, LANGEBAAN

THE MICROSCOPY SOCIETY OF SOUTHERN AFRICA

www.microscopy.co.za

The aim of the Society is to promote and develop microscopy and associated techniques at all levels in the South African region.

The Society was founded as the Electron Microscopy Society of Southern Africa (EMSSA) in 1962 and became a formal member of IFSM (the International Federation of Societies for Microscopy) in 1966. The Society changed its name to the Microscopy Society of Southern Africa (MSSA) in 1996 to accommodate all forms of microscopy.

The primary objectives of the Society are:

- To further microscopy in the broadest sense all branches of science,
- To facilitate communication and co-operation between microscopists and with other scientists,
- To act as a liaison between members of the Society and IFSM,
- To provide an interface between research, education, government and the public.

This is achieved mainly through:

- an annual meeting of the Society,
- publication of a Newsletter,
- maintenance of an Instrument and Skills database,
- the establishment of regional groups to promote closer co-operative ties.
- an email list server: massa@kendy.up.ac.za

Honorary Members:

Prof B I Balinsky, Prof P Berjak, Prof J Coetzee, Dr N R Comins, Mr R H M Cross, Prof JAA Engelbrecht, Dr J T, Fourie, Mr D Geppert, Mr A N Hall, Prof ME Lee, Dr J Murphy, Prof F Nabarro, Prof JH Neethling, Prof B T Sewell, Prof J T Soley, Dr E M Veenstra, Prof T A Villiers, Dr J Wesley-Smith, Prof H Wilsdorf, Prof M J Witcomb, Dr S Wolfe-Coote.

Cover Micrograph: Laves Phase in as Cast Slab of Type AISI 436 Stainless Steel

Image taken by Jaco Kruger from Columbus Stainless Pty Ltd

MICROSCOPY SOCIETY OF SOUTHERN AFRICA

MIKROSKOPIEVERENIGING VAN SUIDELIKE AFRIKA

UMBUTHO WABAKOPOLI BAMAZANTSI E-AFRIKA

PROCEEDINGS
VERRIGTINGE
INKQUBO-NGXOXO

VOLUME, UMQULU 48, 2019

CLUB MYKONOS, LANGEBAAN

The 56th Annual Conference was held at Club Mykonos, Langebaan 1 to 5 December 2019.

This volume contains the Proceedings of the Microscopy Society of Southern Africa and is published annually to coincide with the conferences of the Society. The Society does not hold itself responsible for errors in the manuscripts and does not necessarily subscribe to the opinions and conclusions of the authors.

Die 56ste Jaarlikse Konferensie het by Club Mykonos, Langebaan, plaasgevind vanaf 1 – 5 Desember 2019.

Hierdie volume bevat die Verrigtinge van die Mikroskopievereniging van Suidelike Afrika en word uitgegee om met die konferensies van die Vereniging saam te val. Die Vereniging aanvaar nie verantwoordelikeheid vir foute in die manuskripte nie en stem ook nie noodwendig met die menings en gevolgtrekkings van die outeurs saam nie.

Ingqungquthela yeminyaka yama 56 eyayibanjwelwe kwiziko lengqunqguthela lase Club Mykonos, Langebaan, kwisixeko sase Bela-Bela ukusuka kumhla we 1 ukuyokuma kowe 5 kweyoMnga ngonyaka wama 2019.

Lomqulu uqulethe inkqubo-ngxoxo zendibano yequmrhu leziKopoli zase maZantsi e-Afrika epapashwa ngonyaka ukuze ingqamane nale ngqungquthela yeliqumrhu. Eliqumrhu liyazikhulula kwiimpazamo ezingakho kwiingqokelela zeentetho ezikulo mqulu kwaye ungarhumi nezimvo nezigqibo zababhali.

EDITORS/REDAKTEURS/ABAHLELI

The editors wish to acknowledge the assistance of the many members who refereed the papers printed in the Proceedings.

Die redakteurs wil graag hul dank uitspreek teenoor die vele lede wat die bydraes in die verrigtinge beoordeel het.

Abahleli banqwenelela ukukhankanya inkxaso yamalungu ngobuninzi babo abathathe inxaxheba ekushicilelweni kwala maphepha alenkqubo.

EXECUTIVE COMMITTEE 2019/UITVOERENDE KOMITEE 2019/ ISIGQEBA SOLAWULO 2019

Published by the Microscopy Society of Southern Africa, c/o Council for Scientific and Industrial Research, 1-Meiring Naude Road, Brummeria , Pretoria, 0001, South Africa

ISSN 0250-0418

MICROSCOPY SOCIETY OF SOUTHERN AFRICA
FIFTY SIXTH ANNUAL CONFERENCE: 2019

Table of Contents

Invited Lectures

Life Sciences

Materials Sciences

Materials Sciences Workshop

RECENT ADVANCES IN SIMULATION-ASSISTED ELECTRON BACKSCATTER DIFFRACTION

M. De Graef

Department of Materials Science and Engineering, Carnegie Mellon University, Pittsburgh, USA

Electron backscatter diffraction (EBSD) has been a staple characterization technique for the materials science and geology communities. Since its commercial introduction in the early 1990s, detector systems have steadily improved, and today's systems are capable of acquiring and indexing several thousand patterns per second. The indexing algorithm, which is based on the Hough transform to locate the Kikuchi bands, has not evolved very much since its introduction; it is well known that EBSD patterns with low signal-to-noise ratio, or overlap patterns near grain or phase boundaries, pose difficulties for most of the commercial systems. In this contribution, we will review several recent approaches to the indexing of EBSD patterns that are more robust against noise, and that can provide additional information not easily extracted using the traditional indexing approaches.

The rapid advances in computer hardware enabled by Moore's Law and the cheap data storage space made available by modern media have made it possible (1) to generate accurate, fast, and realistic forward models for the prediction of EBSD patterns, and (2) to store *all* the experimental diffraction patterns on disk, which allows for additional analysis after data acquisition is complete.

In contrast to Hough-based indexing (HI), which relies on the extraction of features (i.e., Kikuchi bands) from the experimental patterns, the new indexing approaches do not look for image features but, instead, attempt to match the complete diffraction pattern by comparison with a pre-computed "master EBSD pattern." The master pattern represents the backscattered electron yield on a spherical surface (the Kikuchi sphere) and is computed based on a combination of Monte Carlo and dynamical scattering simulations. From this master pattern, individual EBSD patterns can be computed by means of bilinear interpolation, assuming that the detector geometry has been calibrated accurately. Fig. 1 shows example master patterns for α-Ti (left) and β-Ti (right) for a microscope voltage of 15 kV. Note that the crystallographic [100] direction points towards the right, and the pattern normal is parallel to the normal to the plane formed by the [100] and [010] directions. The master pattern also inherits the symmetry of the crystal lattice.

The Dictionary Indexing (DI) approach[1] employs a uniform sampling of orientation space SO(3) to create a library (dictionary) of predicted patterns. The sampling algorithm generates a uniform set of orientations inside the Rodrigues Fundamental Zone (FZ) for the given crystal symmetry. An example sampling is shown in Fig. 2 for cubic crystal symmetry; each sphere on the right represents a unique orientation in the FZ (left). For each of these orientations, an EBSD pattern is generated by bilinear interpolation from the master pattern. Each experimental pattern is compared to every dictionary pattern using a pattern matching metric, usually the normalized dot product. The library pattern that has the highest similarity metric then provides the orientation of the experimental pattern; this orientation can be refined

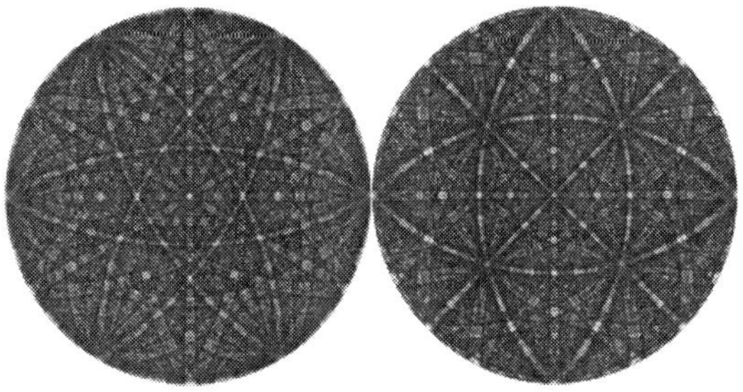

Figure 1. Example master patterns for α-Ti (left) and β-Ti (right) for a microscope voltage of 15 kV.

further, typically to an accuracy of 0.1°. The DI approach is highly robust against noise, as we will demonstrate using a series of Ni data sets as well as a 3.5-million pattern data set from a shot-peened Al sample and a 360-million pattern data set from a Ni-based superalloy. We will also discuss the advantages and drawbacks of the DI technique (which is available to the community in open source form as the *EMsoft* package[2]).

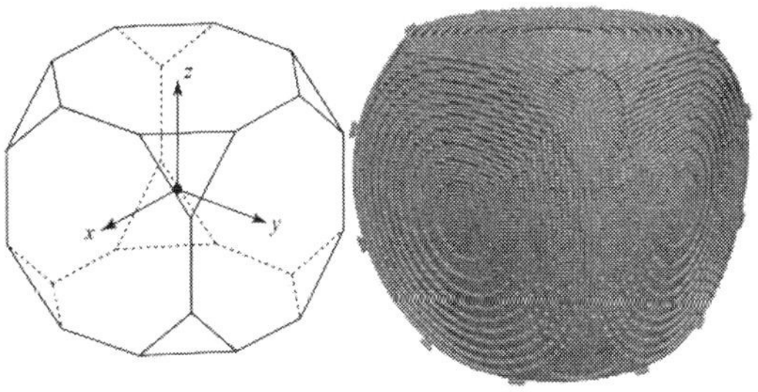

Figure 2. (left) Rodrigues Fundamental Zone for the cubic crystal system; (right) uniform sampling of this zone, with each of the 333,227 spheres representing a unique crystal orientation.

An example of the difference between Hough-based indexing and dictionary indexing is shown in Fig. 3 for the case of shot peened Al at 10 kV. Black pixels on the left are un-indexed points. Note that the smallest grains near the shot-peen crater are dynamically recrystallized grains and they are well resolved by the dictionary indexing approach, even for grains of 150 nm size[3].

A second new indexing approach, Spherical Indexing[4] (SI), is based on the spherical harmonic transform of the master pattern and back-projects the experimental patterns onto the Kikuchi sphere; a cross-correlation on

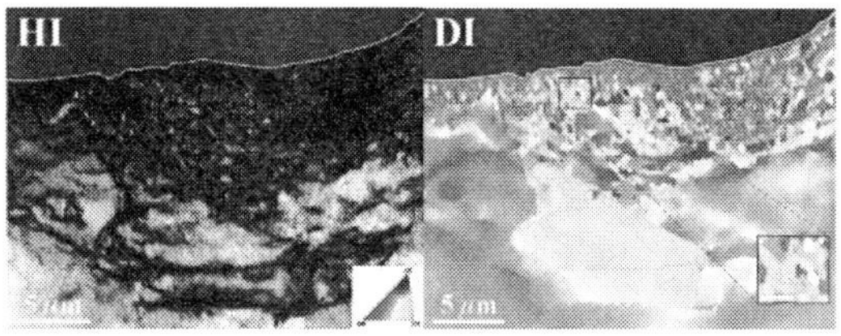

Figure 3. Example indexing results for a shot-peened Al sample: Hough-based indexing on the left (black pixels not indexed) and Dictionary Indexing on the right.

the sphere then directly provides the orientation as an Euler angle triplet. SI is equally robust to noise but is generally much faster than DI since the indexing rate does not depend on the crystal symmetry; for the DI approach, the size of the dictionary increases with decreasing crystal symmetry. We will provide examples of the SI approach and compare it quantitatively to the DI technique in terms of execution speed and accuracy. In particular, we will compare the two approaches using both a Ni data set with high quality patterns, and a Ni-based superalloy data set from a 1,000 layer serial sectioning experiment for which the signal-to-noise ratio of the patterns is relatively low.

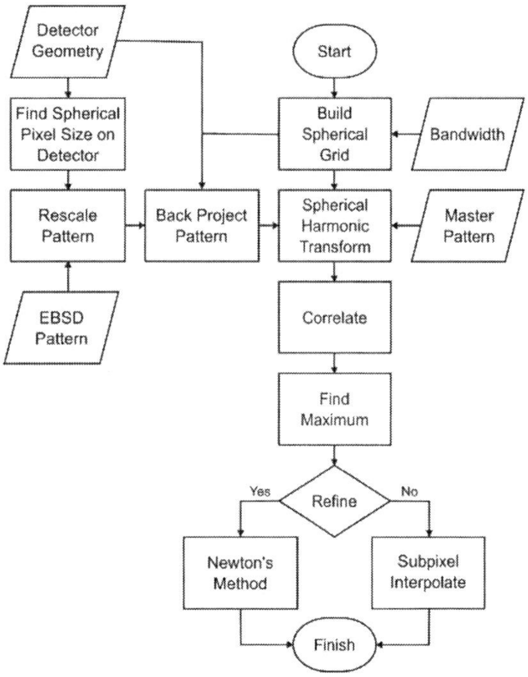

Figure 4. Simplified flow chart for SI approach.

One additional advantage of the SI approach is that it allows for the fast computation of the auto-correlation of the EBSD master pattern. This produces an intensity distribution in Euler space that has primary maxima at the locations of the symmetry elements of the crystal structure, and secondary maxima at orientations that are likely candidates for pseudo-symmetry issues[5] which lead to systematically mis-indexed pixels in the HI approach. The use of the Kikuchi master pattern as the main input to the DI and SI indexing algorithms also allows for unconventional applications; for instance, in cases where the microstructure is very fine, such that multiple phases are present in the beam interaction volume, one can compute a weighted average of the relevant master patterns, rotated into the correct orientation relation, and then use the overlap master pattern to index the experimental patterns. This results in orientation maps for samples that are difficult or impossible to index with other approaches; such maps have been used to obtain the parent phase and variant orientations in $\alpha-\beta$ Ti alloys.

Since the international geological community has a large number of avid EBSD users, we will also discuss the indexing, using both DI and SI approaches, of a 500,000-pattern multi-phase dataset consisting of cubic garnet, monoclinic clinopyroxene, and an amorphous melt phase. This data set provides a nice opportunity for quantitative comparison of DI and SI techniques against the commercial HI approach. A second geological data set for orthorhombic forsterite (Mg_2SiO_6) will be used to illustrate why accurate and robust indexing is important; the random texture of this sample should produce a misorientation plot that corresponds to the theoretical MacKenzie plot for random textures, but, due to pseudo-symmetry effects, the HI approach does not provide the correct result[6]. Both DI and SI approaches correctly index the patterns and avoid errors due to pseudo-symmetry; the resulting MacKenzie plots closely follow the theoretical predictions.

The SI algorithm offers the possibility of real-time pattern indexing. In numerical tests, we have achieved indexing rates approaching 1,500 patterns per second using a standard 24-core Linux work station; a detailed performance analysis revealed that this rate is limited by disk access speeds, reading the patterns from file. If, instead, the patterns were to come directly from a detector, then we anticipate that indexing rates in excess of 5,000 patterns per second should be achievable, keeping up with the fastest new detector systems. Coupling of the acquisition hardware with a high performance computing cluster, with multiple graphical processing units (GPUs) could further increase the indexing rate, thereby freeing up time for additional real-time data analysis. We will conclude this contribution with a brief look towards the future of EBSD.

References
1. Chen, Y.N. *et al.* (2015) Microsc. MicroAnal. 21, 739.
2. https://github.com/EMsoft-org/EMsoft
3. Singh, S. *et al.* (2018) Nature Sci. Rep. 8, 10991.
4. Lenthe, W.C., Singh, S., and De Graef, M. (2019) Ultramicroscopy (in press)
5. Lenthe, W.C., Singh, S., and De Graef, M. (2019) J. Appl. Cryst. 52, 1157.
6. Marquardt, K. *et al.* (2017) American Mineralogist, 102, 1843.

Corresponding author: degraef@cmu.edu

SEEING IS BELIEVING: BUILDING AN IMAGING PLATFORM TO ELUCIDATE CELLULAR AND GENETIC FUNCTION IN *Mycobacterium tuberculosis*

Timothy J. de Wet[1,2,3], Michael A. Reiche[1,2], Ryan Dinkele[1,2], Sophia Gessner[1,2], Zela Martin[1,2,4], Joana A. Santos[5], Saber Anoosheh[1,2], Atondaho Ramudzuli[1,2], Dirk M. Lang[6], Jesse Aaron[7], Teng-Leong Chew[7], Rolf Müller[8,9], Musa M. Mhlanga[3], Neeraj Dhar[4], John D. McKinney[4], Roger Woodgate[10], Meindert H. Lamers[5], Valerie Mizrahi[1,2,11], Digby F. Warner[1,2,11]

[1]SAMRC/NHLS/UCT Molecular Mycobacteriology Research Unit, DST/NRF Centre of Excellence for Biomedical TB Research, Department of Pathology, University of Cape Town, South Africa; [2]Institute of Infectious Disease and Molecular Medicine, University of Cape Town, South Africa; [3]Department of Integrative Biomedical Sciences, University of Cape Town, South Africa; [4]Laboratory of Microbiology and Microsystems, School of Life Sciences, Swiss Federal Institute of Technology in Lausanne (EPFL), Lausanne, Switzerland; [5]Leiden University Medical Center, Netherlands; [5]Confocal and Light Microscope Imaging Facility, Department of Human Biology, University of Cape Town, South Africa; [6]Advanced Imaging Center, Howard Hughes Medical Institute, United States of America; [8]Department of Microbial Natural Products, Helmholtz Institute for Pharmaceutical Research Saarland, Helmholtz Centre for Infection Research and Pharmaceutical Biotechnology, Saarland University, Germany; [9]German Centre for Infection Research (DZIF), Partner Site Hannover-Braunschweig, Germany; [10]Laboratory of Genomic Integrity, National Institute of Child Health and Human Development, National Institutes of Health, United States of America; [11]Wellcome Centre for Infectious Diseases Research in Africa, University of Cape Town, South Africa.

Understanding cellular and genetic function in *Mycobacterium tuberculosis* is critical to the development of new therapeutic approaches (including novel antibiotics) for tuberculosis (TB)[1], currently the leading cause of mortality owing to a single infectious agent and source of one-third of all antimicrobial resistant (AMR) deaths[2]. While an expansive array of genetic and molecular tools are available for mycobacterial manipulation and experimentation, difficulties inherent in working with slow-growing, infectious organisms in access-restricted biological safety facilities have limited the capacity for advanced imaging – especially of live cells – to a small number of laboratories globally. To address this limitation, we have developed a platform equipped for analysis of both infectious and non-infectious material under a growing array of imaging modalities. Here, two examples will be presented illustrating the application of different microscopy techniques to investigate cellular and genetic function in mycobacteria.

The first centres on a mutagenic DNA repair system that has been implicated in DNA damage-induced drug resistance and host adaptation in *M. tuberculosis*[3]. Although the available microbiological evidence is strong, the precise molecular composition and operation of this "mycobacterial mutasome" – minimally comprising DnaE2 polymerase and ImuA′ and ImuB accessory proteins – has remained elusive. Applying a panel of fluorescently tagged translational reporters, we observe that following exposure of mycobacteria to genotoxic stress, ImuB co-localizes with the β clamp in distinct foci whose number increases proportionally with the extent of bacillary filamentation. In contrast, neither DnaE2 nor ImuA′ exhibits discernible subcellular distributions. Notably, functional inactivation of the mutasome in an *imuB*^{AAAAG} mutant containing adisrupted β clamp-binding motif abolishes ImuB-β focus formation, a phenotype recapitulated in bacilli treated with the novel natural product, griselimycin, and in biochemical assays in which this β-binding antibiotic disrupts pre-formed ImuB-β complexes. These observations reinforce the essentiality of the ImuB-β interaction for mutagenic DNA repair in mycobacteria, simultaneously identifying the mutasome as target for compounds designed to protect anti-TB antibiotics against emergent drug resistance.

In the second example, CRISPR interference and quantitative, image-based analyses have been used to construct and characterize an arrayed library of essential gene knockdown mutants in the related non-pathogen, *M. smegmatis*[4]. Applying automated imaging and analysis tools, we derive robust quantitative descriptions of bacterial morphologies, providing an atlas of morphological changes consequent on essential gene-depletion in mycobacteria. Leveraging statistical-learning approaches, we demonstrate that functionally related genes cluster based on morphological similarity, and that bacterial morphology can be used to suggest hypothetical gene function and to infer drug mechanism-of-action.

Our results support the application of large-scale image-based analyses to essential gene function in mycobacteria, and suggest the utility of this approach for drug mechanism-of-action studies as well as the identification of novel targets that might potentiate existing anti-TB agents.

References:
1. de Wet T.J., Warner, D.F. and Mizrahi, V. (2019) *Acc Chem Res*. doi: 10.1021/acs.accounts.9b00275.
2. WHO (2019) Global Tuberculosis Report 2019. Geneva: World Health Organization. Licence: CCBY-NC-SA3.0IGO.
3. Warner D.F. *et al.* (2010) Proc. Natl. Acad. Sci. USA 107, 13093.
4. de Wet, T.J. *et al.* (2018) bioRxiv 358275; doi: 10.1101/358275.
5. Kamariza, M. *et al.* (2018) Sci. Transl. Med. 2018 10(430). pii: eaam6310. doi: 10.1126/scitranslmed.aam6310

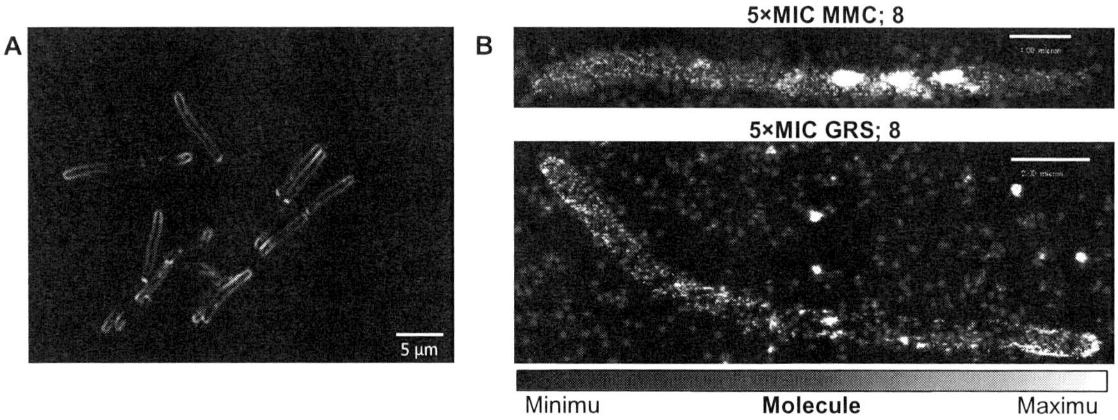

Figure 1. (*A*) SIM image of engineered *M. smegmatis* strain expressing an mCherry-tagged DNA replication protein (Red) and stained with a membrane dye (DMN-trehalose, Green)[5]. Image was captured utilizing a Zeiss Elyra S.1 equipped with an Andor EM-CCD camera iXon DU 897. Scale bar, 5 µm. (*B*) Super-resolution imaging confirms disruption of ImuB-β clamp foci in MEos4A-ImuB mycobacterial reporter treated with griselimycin (GRS). Representative iPALM micrographs of bacilli exposed to 5× MIC mitomycin C (MMC) (top panel) or 5× MIC GRS (bottom panel). Sub-diffraction-limited super-resolution localization of MEos4A-ImuB is observed as highly dense localizations of signal following exposure to MMC; in contrast, GRS prevents the formation of high-density fluorophore signal. Scale bars are 1 µm and 2 µm in the top and bottom micrographs, respectively; molecule probability represents the fluorescence signal probability from minimum (black) to maximum (white) likelihood. iPALM images were captured in collaboration with Drs. Teng-Leong Chew and Jesse Aaron in the Advanced Imaging Center, HHMI Janelia Farm Campus, USA.

3D VISUALIZATION OF HUMAN HAIR WITH VARYING DEGREES OF CURLINESS USING NANO-CT

C.B. van den Berg[1], N.P. Khumalo[1] and M. Ngoepe[2]

[1]Hair and Skin Research Laboratory, University of Cape Town, Cape Town, [2]Dept. of Mechanical Engineering, University of Cape Town, Cape Town

The shape of human scalp hair has a wide variability that ranges from straight, to wavy and very curly. There is significant variation in the shape, curvature and cross-sectional area of the hair fiber[1,2]. Geometric parameters based on the degree of curvature, waviness and number of twists along the hair fiber have shown that hair curliness can be classified into 8-groups[3] or a more reliable 6-groups[4]. The hair fiber is a hierarchical structure comprising of an outer cuticle, middle cortex and an inconsistently present innermost medulla layer[5]. In order to study the morphology of the hair fiber in its natural form, one needs a technique that requires minimal sample preparation (embedding, sectioning, staining etc.), thus to preserve the structure. Nano computed tomography (nano-CT) is a non-destructive method capable of imaging the surface and internal three-dimensional (3D) structure of a solid object[6]. In this study, hair with varying degrees of curliness is investigated using nano-CT. The limit to which the nano-CT can resolve detail of the hair fiber at the microstructural level was investigated.

Three different hair types were selected for this study, namely a low curl (type II), a medium curl (type IV) and a high curl (type VI). All three hair samples had not undergone any chemical treatment with hair dye and/or chemical straighteners or relaxes for at least 7 months. The hair was washed using a pH adjusted solution of 5% SDS (pH 5-6), followed by air-drying at room temperature. The hair was secured in place using a plastic frame before being mounted. Nano-CT scans were performed using a General Electric Phoenix Nanotom S system[6]. Hair samples were scanned at 1µm and 2µm resolution. Commercially available software (ScanIP, Simpleware and Volume Graphics VGStudioMax 3.2) was used for image reconstruction and analysis of the CT data.

The 3D model of the hair fibers (Figs 1 and 2) were used to analyze the cross-sectional area at various points along the fiber axis. It was evident that there were observable differences amongst the three hair types. The cross-sectional area was found to be greater in the low curled hair fiber. The highly curled hair showed more variation in the cross-sectional area compared to the low curled hair. The regions where the curly hair constricts, repeated at regular intervals along the fiber. No such pattern was observed in the low curled hair fiber. The nano-CT images from the high-resolution scan (Fig. 3) did not show any clear distinction between the three regions of the hair fiber. However, the nano-CT images from the low-resolution scan was able to resolve the overlapping cuticle scales on the surface of the hair fiber (inset of Figs 1 and 2). In conclusion, nano-CT was unable to resolve the internal microstructure of the hair fiber but could provide information about the surface detail.

References:
1. Westgate, G.E., Ginger, R.S. and Green, M.R. (2017) Exp. Dermatol. 26(6), 483.
2. Lasisi, T. *et al.* (2016) Am. J. Phys. Anthropol. 160(2), 341.
3. De La Mettrie, R. *et al.* (2007) Hum. Biol. 79(3), 265.
4. Mkentane, K. *et al.* (2017) PloS One 12(6), e0172834.
5. Robbins, C.R. (2012) Chemical and Physical Behavior of Human Hair. Springer Science & Business Media.
6. du Plessis, A., le Roux, S.G. and Guelpa, A. (2016) Nucl. Instrum. Methods Phys. Res., Sect. B. 384, 42.

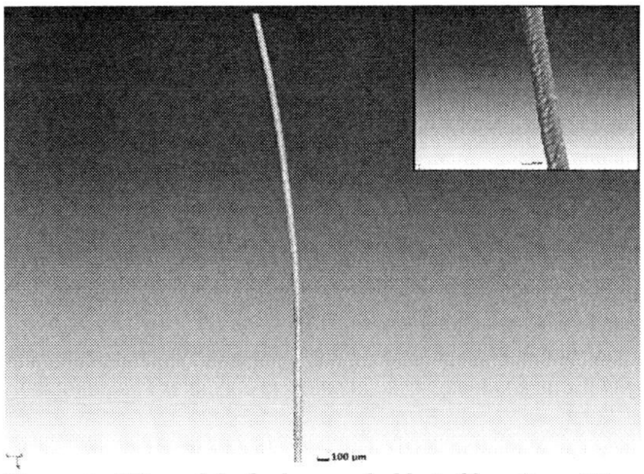

Figure 1. 3D model of a low curled hair fiber (type II).

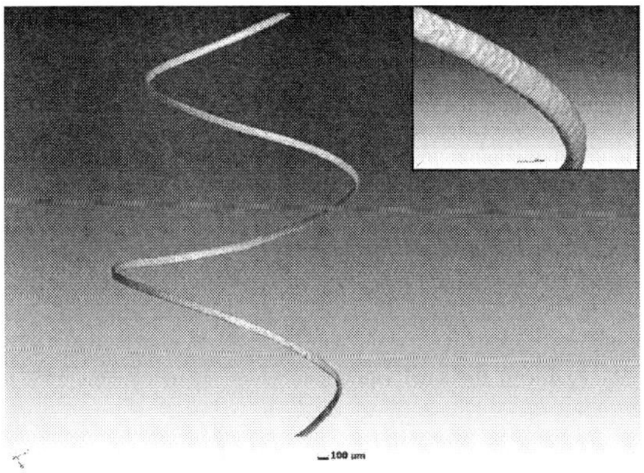

Figure 2. 3D model of a high curled hair fiber (type VI).

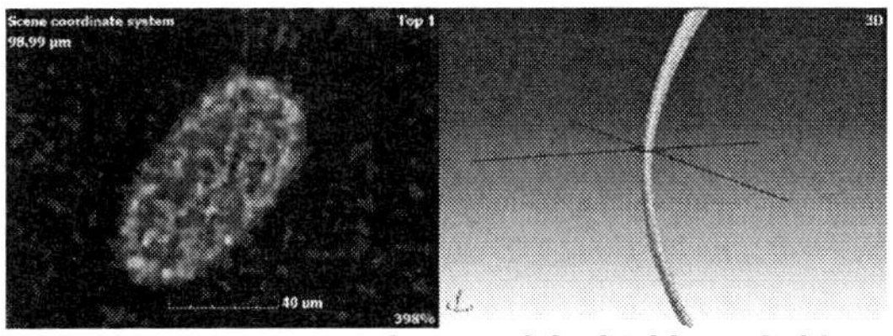

Figure 3. Cross-sectional view of the highly curled hair fiber (left) at the plane indicated by blue cross (right).

Corresponding author: malebogo.ngoepe@uct.ac.za

A WATER-COLLECTING BIOSURFACE IN THE AVIAN NASAL VESTIBULE

M.R. Crole[1], J.H. O'Connell[2], E. Heiss[3], S.J. Werner[4] and H. Rowland[5]

[1]Department of Anatomy and Physiology, Faculty of Veterinary Science, University of Pretoria, Onderstepoort, South Africa, [2]Centre for HRTEM, Nelson Mandela Metropolitan University, Port Elizabeth, South Africa, [3]Institute of Zoology and Evolutionary Research, Friedrich-Schiller-University, Jena, Germany, [4]U.S. Department of Agriculture, Animal and Plant Health Inspection Service, Wildlife Services, National Wildlife Research Center, Fort Collins, Colorado, United States of America, [5]Max Planck Institute for Chemical Ecology, Jena, Germany

Flight can expose birds to environmental and physiological extremes[1]. Low air temperatures and humidity, the need to lose excess body heat[2] and the need to balance energy and water to sustain flight are such examples. Water loss is decreased during nasal breathing and the importance of the nasal cavity for recovering water from exhaled air is widely known[2]. However, despite numerous studies and physiological predictions, the debate continues as to whether energy or water is the limiting factor on flight time[3]. Birds seem to lose less water than predicted by modelling; and after long flights do not appear to be dehydrated[4]. Despite the known function of the respiratory turbinates[5], the avian nasal vestibule has been a neglected area of study.

Five *Sturnus vulgaris,* 1 *Onychognathus morio,* 6 *Acridotheres tristis* and 5 *Agelius phoeniceus* heads were immersion-fixed in 10% neutral-buffered formalin. The nasal cavity was dissected and imaged with a stereomicroscope. Decalcification was done either with 8% formic acid over 6 weeks or with Osteomoll over 4 days. The nasal cavity was embedded in paraffin wax, the blocks sectioned and routinely processed for light microscopy and stained with H&E, PAS and Mallory's Azan.

The nasal vestibule was divided by the nasal septum and occupied by the paired rostral nasal concha (Fig. 1a-c). Additionally, the shortened nasal vestibule in *A. phoeniceus* displayed interlocking horizontal vanes from the lateral vestibular wall, the concha and nasal septum (Fig. 1c) and a surface adorned with macrobumps (±0,25mm). The entire nasal vestibule and structures therein was lined by a modified keratinised, stratified squamous epithelium which in transverse section appeared corrugated (Fig. 1d) and on surface view was cobbled in a hexagonal pattern with a raised centre (Fig. 1e). The dermal-epidermal junction was smooth and the surface peaks were thus formed by the nuclei (which maintained their size to the *Stratum corneum*) of the *Str. intermedium* arranged into vertical columns (Fig. 1d). The epithelium thus formed a vast surface of microbumps about 10µm high.

The increase of surface area of the nasal vestibule is achieved by the rostral nasal concha, the various elaborations and macrobumps as seen in *A. phoeniceus,* and the epithelium of microbumps. Moisture harvesting in arid environments is well known and understood in many reptiles[6] and beetles[7]. However, despite birds living in many extreme environments, moisture harvesting is not known for any non-feathered epidermal structures. A hypothetical surface describes bumps with a small radius and an asymmetric slope as having the ability to form larger condensation drops under an unfavourable temperature gradient and transport water independent of gravity[8]. These features, also present in the avian nasal vestibular epithelium, most likely represent a water-collecting biosurface in birds. Future studies will mathematically quantify the physical properties of this surface, focus on ultrastructure and determine how water transport occurs.

References:

1. Maina, J.N. (2000) J. Exp. Biol. 203, 3045.
2. Giladi, I. and Pinshow, B. (1999) J. Comp. Physiol. B. 169, 311.
3. Carmi, N. *et al* (1992) Auk. 109, 268.
4. Dawson, W.R. (1982) Comp. Biochem. Physiol. 71, 495.
5. Geist, N.R. (2000) Physiol. Biochem. Zool. 73, 581.
6. Sherbrooke, W.C. (1990) J. Herpetol. 24, 302.
7. Comanns, P. *et al* (2011) J. Nanotechnol. 2, 204.
8. Park, C.-K. *et al* (2016) Nature. 531, 78.

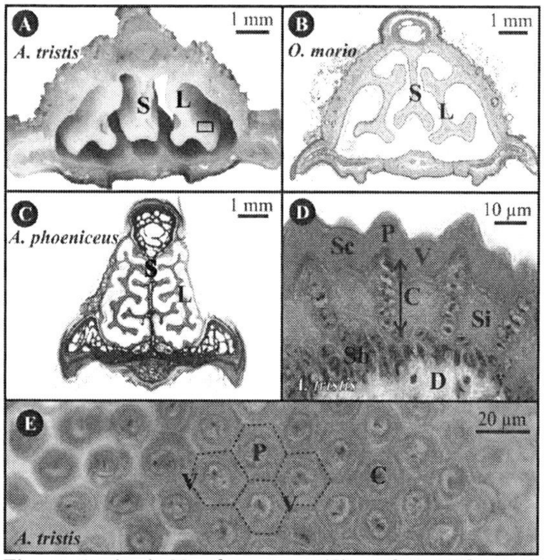

Figure 1. (a-c) Nasal septum (S) and rostral nasal concha (L). Square enlarged in d. (d-e). Epithelium in transverse (d) and surface (e) view. *Str. intermedium* (Si), nuclei columns (C), *Str. corneum* (Sc), peaks (P), valleys (V) and hexagonal pattern (dotted lines) (e). (d). *Str. basale* (Sb) and dermis (D).

Corresponding author: martina.crole@up.ac.za

AHSV NON-STRUCTURAL PROTEIN NS4 INTERFERES WITH JAK-STAT SIGNALING AND SUPPRESSES HOST INNATE IMMUNITY

G.V. Wall[1], A.C. Potgieter[2,3] and V. van Staden[1]

[1]Dept. of Biochemistry, Genetics and Microbiology, University of Pretoria, Pretoria, [2]Deltamune (Pty) Ltd, Centurion, [3]Dept. of Biochemistry, Centre for Human Metabolomics, North-West University, Potchefstroom

African horse sickness virus (AHSV) is a vector-borne pathogen that causes the highly infectious and economically important disease African horse sickness. The virus is non-enveloped and has a genome composed of ten segments of dsRNA, each encoding one viral protein. Genome segment 9, however, encodes NS4 in addition to VP6 from two different reading frames.

Non-structural protein NS4 is the smallest AHSV protein. It is expressed in virus-infected mammalian cells as well as in AHSV-infected horses. The localisation of NS4 to the spleen of infected horses provided an indication that the protein may interact with the host immune system during infection[1]. Early data also suggested that bluetongue virus (BTV) NS4 modulates host innate immunity and the protein has since been shown to be an interferon antagonist[2]. NS4 has been shown to localise to the nucleus of infected cells, which is unusual for a virus that replicates exclusively in the cytoplasm[1].

As a broad aim we are interested in investigating the function of AHSV NS4, and specifically the potential of a role in innate immunity. As a start we used transcriptome sequencing, and analysed differential gene expression (DGE) and KEGG pathway enrichment at 1, 2 or 4 days post-infection in horses infected with virulent AHSV-5 expressing NS4 (rFR) or attenuated AHSV-5 lacking NS4 (rFRΔNS4).

In the horses infected with rFRΔNS4, genes and pathways involved in innate immunity were upregulated on day 1. Fewer genes related to immunity were differentially expressed by day 2, when immunity-related KEGG pathways were no longer enriched or enriched by fewer genes. In contrast, when horses were infected with a virus expressing NS4 (rFR), a similar immune response was only observed on day 2, and continued to day 4.

One of the pathways that stood out in the data was the JAK-STAT signaling pathway, which is particularly important in the interferon (IFN) response central to innate immunity. The products of IFN signaling act as ligands in the activation of the JAK-STAT pathway, leading to the downstream production of antiviral proteins and the initiation of a generalised antiviral state. In order for this to occur, STAT must become phosphorylated and translocate into the nucleus. Some viruses have developed ways to evade the action of IFNs by interfering with this pathway[3]. Thus, we wanted to assess if NS4 affects innate immunity by interfering with the JAK-STAT signaling pathway, in particular with the translocation of STAT into the nucleus. This was done by infecting Vero cells with rFR or rFRΔNS4, followed by IFN-γ treatment at 24 hpi. Normally STAT will translocate into the nucleus after treatment with IFN. Cells were fixed and processed for immunofluorescence with anti-STAT1 primary and AF488 secondary antibodies. Labelling for the AHSV protein NS1 in conjunction with AF594 was used to monitor viral infection and virus gene expression. Slides were viewed with a Zeiss LSM 880 confocal laser scanning microscope coupled to an Airyscan detector.

As expected, STAT1 was observed mainly in the cytoplasm of uninfected or AHSV-infected cells. Upon treatment with IFN-γ, STAT1 translocated into the nucleus of uninfected (not shown) or rFRΔNS4-infected cells (Fig. 1A). In contrast, when rFR-infected cells were treated with IFN-γ, STAT1 remained fully cytoplasmic in the majority of cells and did not translocate to the nucleus (Fig. 1B).

As the only difference between the viruses used in this study is the ability to express NS4, the data presented here provides the first evidence that AHSV NS4 suppresses the host innate immune response in the early stages of infection. One of the mechanisms used may be by interfering with the translocation of STAT1 into the nucleus, so that an antiviral state cannot be achieved.

References:
1. Zwart, L. *et al.* (2015) PLoS ONE 10 (4), e0124281.
2. Ratinier, M. *et al.* (2016) J. Virol. 90 (11), 5427.
3. Fros, J.J. *et al.* (2010) J. Virol. 84 (20), 10877.

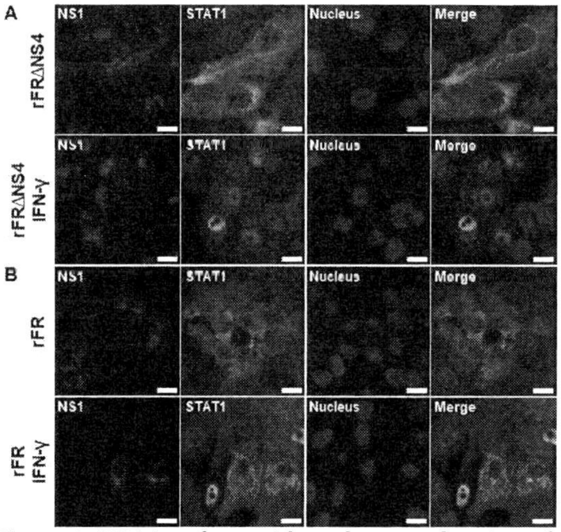

Figure 1. NS4 interferes with nuclear translocation of STAT1. Cells infected with rFRΔNS4 (A) or rFR (B) were untreated or treated with IFN-γ and labelled for NS1 or STAT1. DAPI is shown in blue. Bar = 20 µm.

Corresponding author: gaylevwall@gmail.com

AN ADDITIONAL ROLE OF AHSV VIRUS INCLUSION BODIES AS SITES FOR VIRUS PROTEIN SYNTHESIS

L. Yssel[1], E. Venter[2] and V. van Staden[1]

[1]Dept. of Biochemistry, Genetics & Microbiology, University of Pretoria, [2]Bioimaging, Rothamsted Research, UK

African horse sickness often results in mortality in horses and has a major economic impact in South Africa. The disease is caused by African horse sickness virus (AHSV), which is transmitted by biting midges. AHSV is an orbivirus in the *Reoviridae* family, with a segmented double-stranded RNA genome. Its genome encodes virus structural and non-structural proteins, the latter serving important roles in the viral replication cycle[1].

AHSV and other members of the *Reoviridae* family form dense matrices or "viral factories" in the host cytoplasm, within which new progeny virions are assembled and packaged. In AHSV, these factories are termed virus inclusion bodies (VIBs), and are formed by the aggregation of non-structural protein NS2[2]. In other viruses, VIB-analogues have been shown to interact with host protein synthesis machinery[3].

The aim of this study was to describe the formation of VIBs from NS2 foci in living cells, to determine whether AHSV utilises the VIBs to sequester the cellular machinery to become sites for viral protein synthesis, and to characterise the contribution of NS2 in recruiting specific host translation components to the VIBs.

Two different experimental systems were employed to study NS2 and VIBs. BSR cells were infected with two different serotypes of AHSV or the related bluetongue virus (BTV) to allow the full viral replication cycle to take place. In addition, a transient plasmid-based expression system was used to express NS2 and form VIBs in the absence of other AHSV proteins or viral nucleic acids by transfection with phCMV-NS2 or phCMV-NS2-GFP. This resulted in synthesis of either the native NS2 or an NS2-eGFP fusion protein in BSR cells.

For live-cell imaging, cells were transfected with phCMV-NS2-GFP. To visualise active protein synthesis, virus-infected or phCMV-NS2-transfected cells were subjected to a ribopuromycylation assay, fixed and dual labelled for NS2 and puromycin. To study the interaction between the VIBs and specific host translation proteins, infected or transfected cells were dual labelled for NS2 and either ribosomal components (L11, S3) or translation initiation factors (eIF4E or eIF3θ). All imaging was done with a confocal laser scanning microscope (CLSM, Zeiss LSM 880) coupled to an Airyscan detector.

Live-cell imaging over a 72-hour period revealed that small NS2 foci were highly mobile, and moved in a seemingly random fashion, until separate NS2 foci in close proximity to one another coalesced to form larger aggregates. The ribopuromycylation assay showed that in virus-infected cells, actively translating ribosomes localised to VIBs. However, active translation did not seem to be preferentially localised to VIB-like foci when NS2 was transiently expressed in the absence of other AHSV proteins. The sub-structural localisations of the host translation factors relative to the VIBs in infected or transfected cells showed two different patterns. S3 and eIF4E localised to the periphery of the VIBs, whereas L11 and eIF3θ localized to the central regions. This could be attributed to either biological or technical factors, and should be further investigated.

This is the first attempt to characterise the dynamic nature of AHSV NS2 foci in living, non-fixed cells. This approach could serve as a basis to more accurately describe the processes leading to the formation of AHSV VIBs. This study further provides preliminary evidence that the VIBs are involved in partitioning viral protein synthesis and potentially reveals additional functions of the VIBs. By acting as a hub of virus translation, the VIBs can consequently contribute to suppressing host protein synthesis and elevating the virus life cycle.

References:
1. Roy, P., Mertens, P.P.C. and Casal, I. (1994) Comp. Immunol. Microbiol. Infect. Dis. 17(4), 243.
2. Brookes, S.M., Hyatt, A.D. and Eaton, B.T. (1993) J. Gen. Virol. 74(3), 525.
3. Desmet, E.A., Anguish, L.J. and Parker, J.S.L. (2014) mBio 5(5), 1.

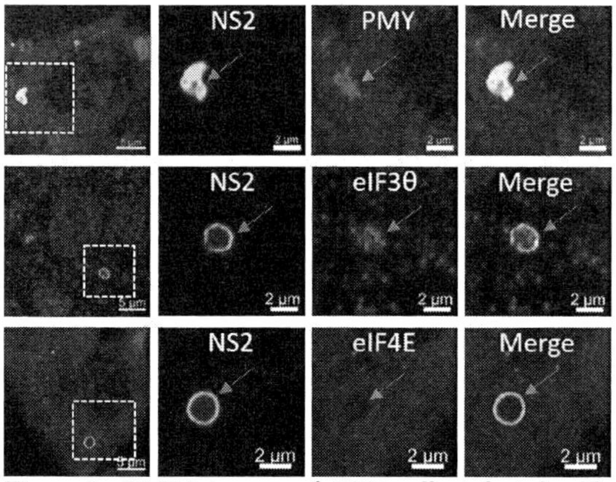

Figure 1. CLSM images of BSR cells infected with AHSV-5. First row shows colocalisation of actively translating ribosomes with the VIBs; middle and last rows show sub-structural localisations for two translation factors within the VIBs. Fluorescent stains: DAPI to show nuclei, anti-NS2/AF488 for VIBs, anti-PMY, anti-eIF3θ and anti-eIF4E/AF594 for protein synthesis and translation factors, respectively. Images to the right are zoomed-in views of the demarked area in the image to the left.

Corresponding author: litiayssel@gmail.com

CHOLECALCIFEROL INDUCES CELL CYCLE ARREST AND APOPTOSIS IN HeLa CELL LINE

Y. Pather [1], S. Marais[1], A.M. Joubert[1], A.E. Mercier[1] and R. Punchoo[2,3]

[1]Department of Physiology, Faculty of Health Sciences, University of Pretoria, [2]National Health Laboratory Services, Tshwane Academic Divison, [3]Department of Chemical Pathology, Faculty of Health Sciences, University of Pretoria

Cervical cancer causes significant morbidity and mortality in South African women. The anticancer actions of vitamin D (VD) compounds (cholecalciferol, calcidiol and calcitriol) have been studied *in vitro* and *in vivo* and results remain inconclusive.[1] Limited studies have investigated VD's anticancer actions in cervical cancer. Cervical cells regulate local VD actions by an intracrine VD metabolizing system. This system includes activating VD precursor enzymes (CYP2R1, CYP27A1 and CYP27B1), an inactivating VD enzyme (CYP24A1) and a VD receptor (VDR).[1,2] This study investigated cholecalciferol's anticancer action and its local regulation of the autocrine VD metabolising system in HeLa cervical adenocarcinoma cell line.

Confluent HeLa cells were treated with cholecalciferol (10 ng/ml, 40 ng/ml, 100 ng/ml and 1000 ng/ml) for 48 hours under standard cell culture conditions. The negative controls included cells propagated in medium-only and equivalent solvent exposure. All experiments were performed in triplicate using three biological repeats. In cytotoxicity studies, cells were enumerated by the crystal violet assay. Cell viability was assessed by the semi-quantitative trypan blue exclusion assay and quantitatively by the 3-(4,5-dimethylthiazol-2-yl)-2,5-diphenyltetrazolium bromide (MTT) assay. Flow cytometric Muse^{TM} assays were used to analyse cell cycle progression; the induction of apoptosis quantified by the caspase 3/7 and Annexin/PI assays; cell proliferation by the Ki67 assay; and autophagic cell death by the LC3-II assay. The lactate dehydrogenase assay evaluated necrotic cell death. Gross and ultrastructural morphological analyses of cell growth and cell death were performed using brightfield and transmission electron microscopy, respectively. In addition, RT-qPCR and Western blots assessed the intracellular VD metabolizing system at mRNA and protein expression levels, respectively. Statistical analysis was performed using ANOVA and post hoc Bonferroni testing ($p < 0.05$ was considered statistically significant). Morphology was assessed qualitatively for features of cell death.

A statistical decrease in HeLa cell count, proliferation and viability; and a statistical increase in cells within the sub-G_0/G_1 and G_0/G_1 phases were identified at 100 ng/ml and 1000 ng/ml cholecalciferol treatments in comparison to negative controls. Statistical increases in Annexin/PI staining and terminal caspases 3/7 activation at 100 ng/ml and 1000 ng/ml treatments in comparison to negative controls and low dose treatments (10 ng/ml and 40 ng/ml) were observed. Morphological hallmarks of apoptosis in experimental treatments (100 ng/ml and 1000 ng/ml) included apoptotic bodies, fragmented nuclei and swollen mitochondria (Fig. 1). Autophagic and necrotic cell death mechanisms were not identified as pathways induced by cholecalciferol treatment. CYP2R1, CYP27B1, CYP24A1 and mitochondrial CYP27A1 mRNA and protein expression levels were significantly increased after 100 ng/ml cholecalciferol treatment. VDR mRNA and protein expressions were also significantly increased in response to 100 ng/ml cholecalciferol treatment. Activated cholecalciferol regulates the expression of genes affecting cell proliferation and cell death via intracrine signalling. Increased CYP2R1 and CYP27A1 expression can activate cholecalciferol to calcidiol. Increased CYP27B1 expression can further activate calcidol to calcitriol.

This study supports the theory of an intracrine activation of cholecalciferol to calcidiol and calcitriol by the intracellular VD metabolising system which induce caspase-dependent apoptosis and G_0/G_1 cell cycle arrest in HeLa cells. Future studies will investigate converging links between disrupted mitochondrial function and megamitochondria formation, free radical species generation and apoptosis. More broadly, this study supports clinical investigation of VD deficiency as a potential risk factor in cervical dysplasia and an adjunctive cancer treatment.

References:
1. Feldman, D., Pike, W. and Adams, J. (2011) Vitamin D. California, Academic Press.
2. Deuster, E. *et al.* (2017) Int. J. Mol. Sci. 18(11), 2328.

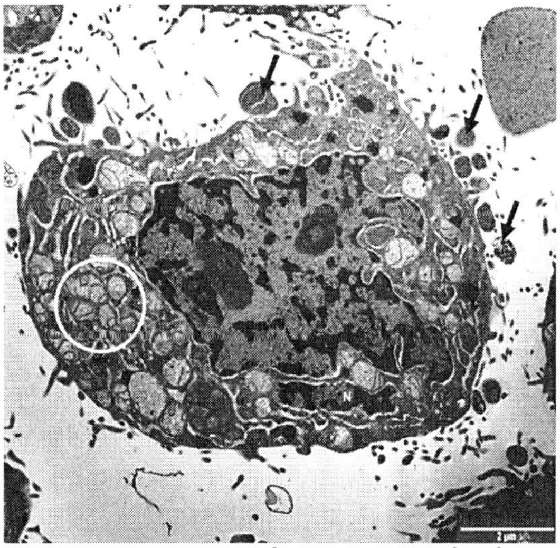

Figure 1. Transmission electron micrograph of 100 ng/ml cholecalciferol treatment of HeLa culture. Cells showed swollen megamitochondria with dilated cristae (encircled), lobed nuclei (N) and apoptotic bodies (arrowed). Scale bar = 2µm.

Corresponding author: rivak.punchoo@up.ac.za

CHOLECALCIFEROL INHIBITS CELL GROWTH AND INDUCES APOPTOSIS IN SiHa CELLS

S. Bhoora[1], T.S. Pillay[1,2] and R. Punchoo[1,2]

[1]Department of Chemical Pathology, Faculty of Health Sciences, University of Pretoria, [2]National Health Laboratory Services, Tshwane Academic Division, Pretoria

Cervical cancer is the fourth most common female malignancy worldwide and is disproportionately higher in low-income countries. In South Africa, cervical cancer is a leading cause of mortality in women.[1] Studies have demonstrated that vitamin D exerts anti-cancer actions in breast, colon and prostate cancer cells.[2] These actions are mediated by calcitriol binding to vitamin D receptor (VDR) and regulating expression of genes responsible for cell growth and differentiation. Vitamin D anti-cancer regulation results in decreased cell proliferation, cell cycle arrest and the induction of apoptosis.[2] It is noteworthy that cervical tissue expresses a localised vitamin D metabolising system of proteins (activating and inactivating vitamin D enzymes and VDR). This system implicates vitamin D in autocrine regulation of growth and differentiation. There is a paucity of studies investigating the anti-cancer action of vitamin D on cervical cancer.[3]

This study investigated the action of precursor vitamin D, cholecalciferol, on cell growth parameters and apoptotic cell death induction in a cervical cancer cell line, SiHa.

Control and experimental SiHa cell cultures were seeded in triplicate, using three biological repeats, and incubated for 72 hours under standard conditions. Controls comprised medium-only and ethanol solvent (0.1%). The cholecalciferol treatment range for experimental cultures was 10 ng/ml, 40 ng/ml, 100 ng/ml and 1000 ng/ml. Cell viability was assessed by the semi-quantitative trypan blue dye exclusion assay. Cell count was enumerated by spectrophotometry with the crystal violet assay which quantified cell number. Cell proliferation was evaluated by flow cytometry using the Muse™ Ki67 cell proliferation assay which measured the percentage of cells staining Ki67-positive indicative of active cell proliferation. Transmission electron microscopy (TEM) was used to study ultrastructural features of cell death. Cells were pelleted, fixed and stained by conventional techniques. Ultrastructure was qualitatively assessed for morphological features of cell death. Statistical analyses were performed using a one-way ANOVA and Bonferroni *post hoc* test using Graphpad Prism (v7., USA).

Cell viability by the trypan blue dye exclusion assay was significantly decreased to 71.8% at 1000 ng/ml and 83.1% at 100 ng/ml treatments in comparison to 90.8% in the medium control; 91.2% in the solvent control; 89.0% at the 10 ng/ml and 90.3% at 40 ng/ml ($p < 0.0001$). Crystal violet studies demonstrated significant decreases in cell number to 62.9% at 1000 ng/ml, 65.1% at 100 ng/ml and 63.7% at 40 ng/ml in comparison to 94.8% in the medium control and 91.3% in the solvent control. Cell proliferation using the Muse™ Ki67 assay was significantly decreased to 80.9% at 1000 ng/ml cholecalciferol in comparison to 94.4% in the medium control, 94.6% in the solvent control, 91.3% at the 10 ng/ml treatment concentration and 90.2% at the 40 ng/ml treatment concentration. Ultrastructural analysis at treatment concentrations of 100 ng/ml and 1000 ng/ml in comparison to controls and low dose treatments (10 ng/ml and 40 ng/ml) showed classical features of apoptosis which included nuclear fragmentation, cell membrane blebbing and apoptotic body formation (Fig. 1).

This study has demonstrated significant inhibition of cell viability, cell count and cell proliferation with associated morphological features of apoptosis at 100 ng/ml and 1000 ng/ml cholecalciferol treatments in comparison to control cultures and low dose treatments (10 ng/ml and 40 ng/ml). These findings are consistent with the anti-cancerous action of vitamin D, which include the inhibition of cell proliferation, and apoptotic induction.[3] Further experiments are being conducted to confirm biochemical apoptosis and elucidate the intracellular regulation of the vitamin D metabolising system in cholecalciferol-treated SiHa cultures.

References:
1. Cohen, P.A. *et al.* (2019) The Lancet 393(10167), 169.
2. Holick, M.F. (2007) N. Engl. J. Med. 357(3), 266.
3. Krishnan, A.V. *et al.* (2014) Nat. Rev. Cancer. 14(5), 342.

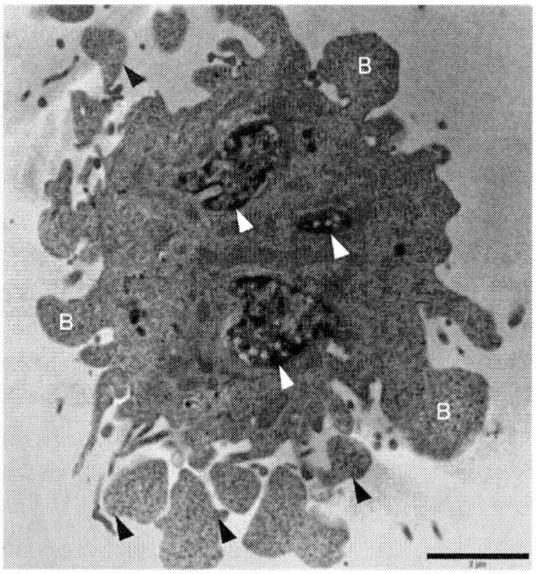

Figure 1. Transmission electron micrograph of SiHa cells treated with 1000 ng/ml cholecalciferol. Morphological features of apoptosis were present and included cell membrane blebbing (B), apoptotic bodies (black arrowheads) and nuclear fragmentation (white arrowheads). Scale bar = 2μm.

Corresponding author: rivak.punchoo@up.ac.za

CILIA IN MOBILINE CILIATES FROM CHITONS: ANOMALIES OR MERELY ODD DEVIATIONS?

L. Basson[1], G.P. de Jager[1] and H. Grobler[2]

[1]Department of Zoology and Entomology, University of the Free State, Bloemfontein, [2]Centre of Microscopy, University of the Free State, Bloemfontein

Mobiline ciliates (Peritricha), represented by three families, are found associated with a variety of invertebrate and vertebrate hosts from freshwater, marine and terrestrial habitats[1]. Representatives of two of these families, i.e. Urceolariidae and Trichodinidae have been studied in southern Africa from various hosts.

Mobilines contain two permanent zones of cilia, the adoral spiral used in food collection and the aboral ciliary girdle used for movement. The adoral spiral consists of two rows, the haplo- and polikineties that both spiral anti-clockwise towards the infundibulum. On the aboral side is the compound ciliary girdle, with a varying number of ciliary rows in the wreath. Most mobiline species studied so far have between 6 – 15 rows, with the inner rows containing shorter cilia, while the cilia become progressively longer towards the outer rows. All these cilia, i.e. the adoral cilia, compound girdle and a row of single marginal cilia present in some species, show the typical elongated cilia structure; long, thin cilia with slightly rounded tips. This condition has been reported for mobilines from freshwater hosts (fish, mollusks and anurans), as well as from marine invertebrate and vertebrate hosts. However, recently mobilines were found on the gills of marine chitons with some peculiarities in cilia structure.

The mobilines were collected from the gills of various species of marine chitons from De Hoop Nature Reserve and prepared for scanning electron microscopy using two different methods. Some were fixed in 10% buffered neutral formalin before washed in tap water. Other samples were fixed in 2% buffered glutaraldehyde at 4°C and then washed in phosphate buffer. Standard procedures for both methods were followed, all specimens dehydrated in ethanol and critical point dried before mounted on aluminium stubs, sputter coated with gold and examined at 5 kV in a JEOL JSM-7800F Field Emission scanning electron microscope.

Various mobiline species studied to date have shown the normal and typical structure in cilia morphology. Freshwater trichodinids (Trichodinidae) found on the skin of freshwater fish showed the usual cilia structure, where the cilia were long and slender, tapering imperceptibly to a slightly thinner point in both adoral and aboral cilia. The same structure was found for trichodinids in the bladder of anurans, i.e. *Trichodina xenopodos* from the clawed toad *Xenopus laevis laevis*[2] and *T. dampanula* from the guttural toad *Sclerophrys gutturalis*[3]. Various species of marine trichodinids from intertidal fish also showed similar and therefore seemingly typical cilia morphology (Fig. 1).

Several populations of *Leiotrocha* (Urceolariidae) collected from the gills of three species of marine chitons, showed some typical cilia. However, certain cilia strongly deviated from the usual and standard cilia reported in mobilines thus far. Cilia in the adoral spiral and ciliary girdle showed both some typical and several irregular cilia. The pattern appeared to be exactly similar in several species of *Leiotrocha* from chitons.

Some of the cilia in the adoral spiral showed regular cilia interspaced between cilia with enlarged flattened spatulate areas near the tip. The ciliary girdle showed an arrangement of typical cilia in the inner shorter cilia. Some of the slightly longer cilia had somewhat rounded tips, some others with slightly rounded tips. The longest outer rows of cilia showed the same anomalous cilia as found in the adoral spiral, where these longer outer girdle cilia had the same widely flattened spatulate areas, again located close to the tip (Figs. 2-3).

It is unclear whether this condition represents a preparation artifact, some odd aberration or whether this represents the "normal" state for cilia of leiotrochans from chitons. Further comparative morphological studies incorporating other invertebrate marine hosts will hopefully answer this perplexing quirk.

References:
1. Basson, L. and van As, J.G. (2006) Trichodinidae and other ciliophorans (Phylum Ciliophora). In: Woo, P.T.K. Fish Diseases and Disorders, Vol 1 Protozoan and Metazoan Infections. Oxfordshire, Cab International.
2. Kruger, J., Basson, L. and van As, J.G. (1991) Syst. Parasitol. 19, 43.
3. van der Bank, M., Basson, L. and van As, J.G. (1989) S. Afr. J. Zool. 24(4), 270.

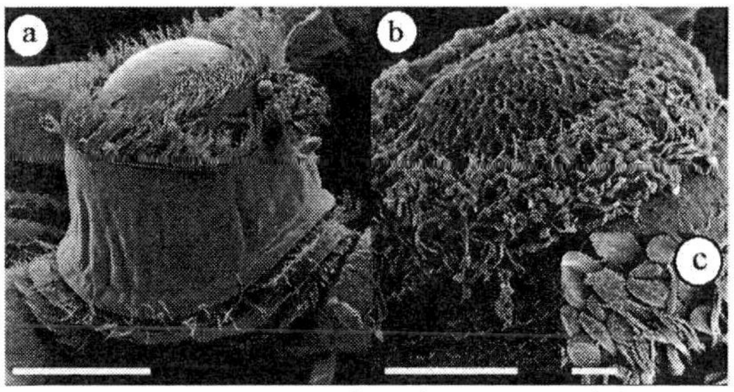

Figure 1. SEM-micrographs of various mobilines. a. Marine trichodinid from *Clinus cottoides* gills with typical cilia. b and c. *Leiotrocha* sp. from gills of the chiton, *Chiton politus* showing irregular cilia (c insert – a close-up of cilia in ciliary girdle). Scale bar = 15 µm (a and b), 0.5 µm (c).

Corresponding author: bassonl@ufs.ac.za

COLOCALISATION OF AFRICAN HORSE SICKNESS VIRUS NON-STRUCTURAL PROTEIN NS4 WITH CELLULAR AND VIRAL PROTEINS

S. Boughan[1], G.V. Wall[1], A.C. Potgieter[2,3] and V. van Staden[1]

[1]Dept. of Biochemistry, Genetics and Microbiology, University of Pretoria, Pretoria, [2]Deltamune (Pty) Ltd, Pretoria, [3]Dept. of Biochemistry, Centre for Human Metabolomics, North-West University, Potchefstroom

African horse sickness (AHS) is a devastating disease in Sub-Saharan Africa, causing up to 95% mortality in susceptible horses. AHS is caused by African horse sickness virus (AHSV), with the virus transmitted via *Culicoides* biting midges. The AHSV virion is a non-enveloped icosahedral structure comprised of seven structural proteins (VP1-VP7), enclosing ten double-stranded RNA (dsRNA) genome segments. These encode all the structural plus four non-structural (NS1-NS4) proteins. Non-structural proteins do not form part of the virion, but play a role in virus replication.

The NS4 protein is encoded by segment 9, which also codes for AHSV VP6[1]. Across all AHSV serotypes, NS4 is categorized into two types (NS4-I and NS4-II) based on size and amino acid sequence. A variant of NS4-II, termed NLS-NS4-II, contains an additional N-terminal nuclear localization signal (NLS). AHSV replicates exclusively in the cytoplasm, however NS4 shows a predominantly nuclear distribution in virus-infected cells in tissue culture. Strong nuclear labelling of NS4 was also detected in stellate-shaped dendritic cells in the spleen of infected horses, indicating that NS4 may play a role in the immune response[2]. Virulent AHSV strains engineered to not express NS4 become attenuated in horses. NS4 of the related bluetongue virus acts as an interferon antagonist, therefore suppressing the immune response upon infection and acting as a virulence factor[3].

The role of AHSV NS4 in the nucleus is not understood. Multiple viruses interact with components of specific nuclear compartments such as promyelocytic (PML) bodies, cajal bodies, nuclear speckles or the nucleolus to give the virus a replication advantage. Viruses can inhibit the formation of these nuclear compartments, or cause them to relocate to the cytoplasm. Different non-structural proteins of a specific virus can also interact or co-function to advance the viral life cycle. In an attempt to establish the function of NS4 in the AHSV replication cycle, the aim of this study was to investigate the interaction of NS4 with host nuclear compartments or other non-structural AHSV proteins.

Mammalian cells were infected with different AHSV strains expressing NS4-I, NS4-II or NLS-NS4-II. Two viruses rescued via reverse genetics were also used, i.e. AHSV-5FR and AHSV-5FR lacking NS4 expression (AHSV-5FRΔNS4). Cells were prepared for confocal laser scanning microscopy (CLSM) by fixing and labelling with antibodies directed against specific nuclear markers (cajal bodies, PML bodies, nuclear speckles) or AHSV non-structural proteins (NS1, NS2, NS3, NS4). Immunofluorescence was viewed using a Zeiss LSM880 confocal microscope with Airyscan detector for super resolution microscopy. Intensity plot profiles to confirm colocalisation was generated using Zeiss ZEN software.

AHSV infection resulted in the relocation of SC35, a protein component of nuclear speckles, to cytoplasmic virus inclusion bodies. Inside the nucleus NS4 did not colocalise with coilin (a marker of cajal bodies), however NS4 did colocalize with PML bodies (Figure 1A). Interestingly, AHSV-5FRΔNS4 infection caused PML bodies to form long tubule-like structures rather than small punctate foci in the nucleus. These tubule-like structures were similar in conformation to the staining profile normally observed for AHSV NS1 tubules. Co-labelling revealed that NS4 did not interact with AHSV NS2 or NS3, however it did co-localize with a small fraction of the NS1 signal that was present in the nucleus. NS1 also colocalised with PML in the nucleus (Figure 1B).

These results indicate a potential interaction of AHSV NS4 and NS1 with components of nuclear PML bodies. As PML bodies play a role in the immune response upon virus infection, this is the first indication of a mechanism whereby AHSV proteins may suppress the host immune response. Understanding the virulence factor NS4 may contribute to AHSV vaccine development.

References:
1. Firth, A. E. (2008) Virol J. 5, 48.
2. Zwart, L. *et al.* (2015) PLoS One 10, e0124281.
3. Ratinier, M. *et al.* (2016) Virol J. 90 (11), 5427.

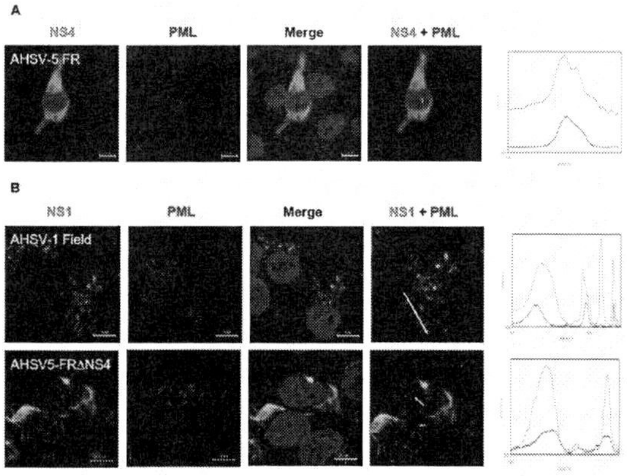

Figure 1. AHSV-infected Hela cells were fixed at 24 hours post infection and dual labelled with rabbit anti-NS4 (A) or anti-NS1 (B) and mouse anti-PML as indicated. Alexa Fluor 488 anti-rabbit or 594 anti-mouse was used as secondary antibody. Nuclei were labelled with DAPI. Scale bars represent 10µM. Intensity profiles are indicated as histograms.

Corresponding author: shareen16@gmail.com

CONSIDERING HEMORHEOLOGY AS A KEY FACTOR IN PROSTATE CANCER MANAGEMENT

J. Bester[1], S. de Villiers[1], A.C. Swanepoel[1] and K. Ramsamy[2]

[1]Department of Physiology, University of Pretoria, Pretoria, [2]Department of Urology, University of Pretoria, Pretoria

Prostate cancer (PCa) is defined as an uncontrolled abnormal cell growth of the prostate gland with malignant characteristics. It is the second most diagnosed cancer in men with an estimated 1.1 million diagnoses in 2012 alone[1]. There have been many years of research into prostate cancer aetiology and a possible cure, yet little research has focused on the management of the cardiovascular complications associated with PCa[2]. There is mounting evidence that the cancer itself and its therapy are associated with a higher risk of venous thromboembolic events[2]. Hemorheology is one of the key factors in identifying hypercoagulability in chronic inflammatory diseases such as PCa and has not been studied in detail in this patient population[3]. Therefore, it can be said that the prothrombotic state of this patient population has not yet fully been studied and accordingly there is a need to understand the hemorheology of PCa patients to develop better management strategies. Therefore, the aims of this study were to investigate whether venous thromboembolic events (VTE) prophylaxis should be considered in these patients prior to introducing therapy, as well as useful ways of profiling the hypercoagulable state of patients by examining their rheology.

A total of 67 treatment naïve PCa patients with no other co-morbidities were recruited over a period of 6 months. A combination of viscoelastic and ultrastructural properties of whole blood in PCa patients were used to study their hemorheology. Scanning electron microscopy (SEM) was used to study the fine morphological changes such as shape changes and membrane topography. A thromboelastograph (TEG®) was used to measure the viscoelasticity during clot formation. The sample population was divided into non-metastatic and metastatic groups.

The TEG® results showed significant differences in the metastatic *vs.* non-metastatic groups (Table 1). The changes seen in the viscoelastic tests suggest stronger, more rigid clots are formed in the metastatic group. Also, the presence of abnormal ultrastructure of the fibrin fibres and platelets was noted in the SEM analysis and confirmed the increased viscosity trends observed in the metastatic group. Both the non-metastatic and metastatic groups had an abnormal morphology, specifically the fibrin fibers, but the changes seen in the metastatic group (Fig. 2) were much more severe compared to the non-metastatic group (Fig. 1). The results obtained provide sufficient evidence to conclude that it is necessary to monitor PCa patients, specifically metastatic patients, much closer for thrombotic events, and to consider VTE prophylaxis before and during treatment.

References:
1. Bray F. *et al.* (2018) CA: A Cancer J. Clin. 68(6), 39.
2. Lima, LG. and Monteiro, RQ. (2013) Biosci. Rep. 33
3. Secin FP. *et al.* (2008) Eur. Urol. 53(1), 134.

Table 1. Summary statistics for TEG® measurements by metastatic status (n = 67). For a description of the parameter abbreviations see https://www.sciencedirect.com/science/article/pii/S0049384817302736?via%3Dihub#t0005

	Non-Metastatic	Metastatic	Mann-Whitney test
Parameter	**Median (IQR)**	**Median (IQR)**	**p-value**
PSA	16.1 (10.6)	164.4 (821.3)	< 0.001
R	8.4 (3.0)	7.4 (3.2)	0.2789
K	2.8 (1.6)	1.8 (1.5)	**0.0152**
Angle	62.2 (13.4)	70.5 (14.2)	**0.0264**
MA	60.5 (11.5)	64.8 (10)	**0.0117**
G	7.7 (3.4)	9.2 (4.2)	**0.0119**
MRTG	4.8 (3.1)	7.6 (7)	**0.0094**
TMRTG	11.8 (5.4)	10 (5.5)	0.1483
TTG	769.3 (320.9)	937.2 (428.4)	**0.0108**

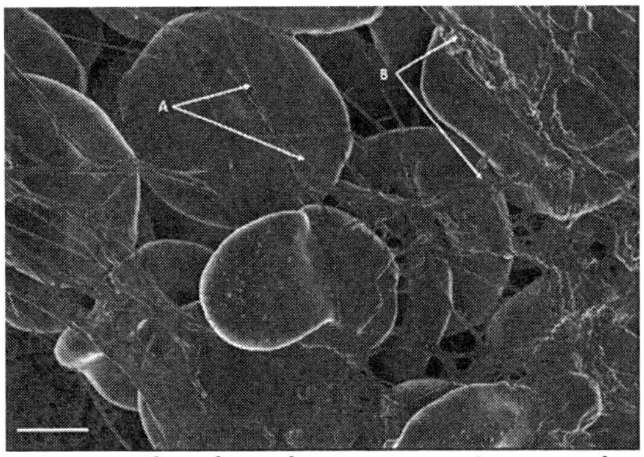

Figure 1. Clots from the non-metastatic group that shows deformed RBCs surrounded by some normal (A) and some abnormal fibrin fibres (B) after thrombin addition. Scale 2μm

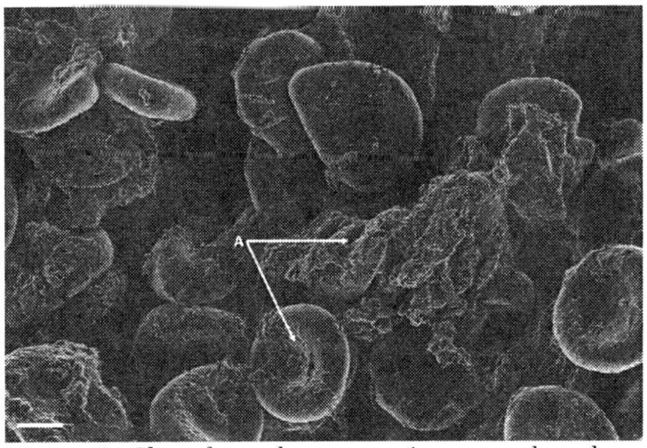

Figure 2. Clots from the metastatic group that show slightly deformed RBCs surrounded by abnormal fibrin deposits and sheaths (A) after thrombin addition. Scale 2μm

Corresponding author: janette.bester@up.ac.za

DIVERSITY OF SESSILINE CILIATES (PERITRICHA: SESSILINA) FROM FISH IN TASMANIA, AUSTRALIA

L. Basson

Department of Zoology and Entomology, University of the Free State, Bloemfontein

Tasmania is the island state of Australia, lying just south of the mainland. The freshwater fish fauna of the island consists of 25 native and eight introduced species. A further 15 species are predominantly marine or estuarine inhabitants that may be encountered some distance inland. Very little information is available on the parasites of these fish, bar three publications[1,2,3].

During four surveys examining the ectosymbionts of fish in Tasmania, several sessiline ciliates were collected. Four fish species hosted a total of nine sessiline species represented by three families and four genera. The family Epistylididae Kahl, 1935 was represented by one species of *Epistylis* Ehrenberg, 1830 and two species of *Apiosoma* Blanchard, 1885. A single species of *Caliperia* Laird, 1953 belonging to the Ellobiophryidae (Chatton and Lwoff, 1929) was found. The last family, i.e. Scyphidiidae Kahl, 1935 was represented by five species of *Ambiphrya* Raabe, 1952.

Fish hosts were collected using seine and scoop nets. Live fish were transported to the laboratory at the University of Tasmania, Hobart, where they were examined for ectosymbionts. Wet smears were made from the skin and gills, examined and if sessiline symbionts were present, the smears were fixed wet in Bouins fixative, after which the smears were transferred to 70% ethanol till staining. Smears were stained with Mayer's Hematoxylin[4]. Small fish were fixed whole in 10% buffered neutral formalin and later processed for scanning electron microscopy by dehydrating in ethanol concentrations, critical point dried, sputter coated and examined in a JEOL JSM-7800F Field Emission scanning electron microscope at 5-10 kV.

A single species of *Epistylis* was found forming small colonies on the skin of the indigenous soldier fish, *Gymnapistes marmoratus* (Cuvier, 1829). This scorpaenid fish is abundant in all southern Australian eelgrass beds. Two species of *Apiosoma* (Fig. 1) and two species of *Ambiphrya* (Figs 2-3) were found on the skin and gills of the introduced red fin perch, *Perca fluviatilis*. The girdled goby, *Nesogobius* sp. 1 is one of the most abundant inhabitants in Tasmanian and Victorian waters and was found to host a single *Ambiphrya* sp. The last fish is known as the Tamar hardyhead, *Leptatherina presbyteroides*, an abundant atherinid that schools in shallow marine environments along the southern New South Wales and around Tasmania. A single species of *Caliperia* was found on the skin of this small fish, as well as two species of *Ambiphrya* from the skin and gills.

The introduced perch showed the highest diversity of sessiline symbionts, some or all of which may have been introduced. While these ciliates are never a problem to healthy fish, other introduced parasites may be problematic.

Sessiline ciliates never pose a threat to their fish hosts and are considered to be true commensals. They are filter feeders, feeding on bacteria and organic debris in the surrounding water. Under natural conditions, sessiline ciliates never occur in large numbers on healthy hosts. An abnormally high occurrence on any host is not a sign of a pathogenic occurrence, but rather an indication of some adverse environmental factor playing a role. Therefore, the occurrence of sessilines is very often used as an indicator of the health of an ecosystem.

Tasmanian fish species are facing a number of problems (such as industrial pollution, removal of feeding, breeding or shelter areas, use of chemicals and land clearing practices). As natural habitats are changed or damaged due to increasing human activities, the endemic and native fish fauna of Tasmania will be under more pressure. The role of introduced parasites must be monitored under these conditions, but in order for that to happen, there must be baseline surveys done of indigenous and introduced fish and their parasites and/or commensals before any meaningful deductions can be made, something regrettably lacking on the island state of Tasmania.

References:
1. Fryer, G. (1969) Austr. J. Zool. 17, 49.
2. Basson, L. (2010) Acta Protozool. 49, 253.
3. Su, X.Q. and White, R.W.G. (1995) Syst. Parasitol. 32, 53.
4. Humason, G.L. (1979) Animal Tissue Techniques. San Francisco, W.H. Freeman and Company.

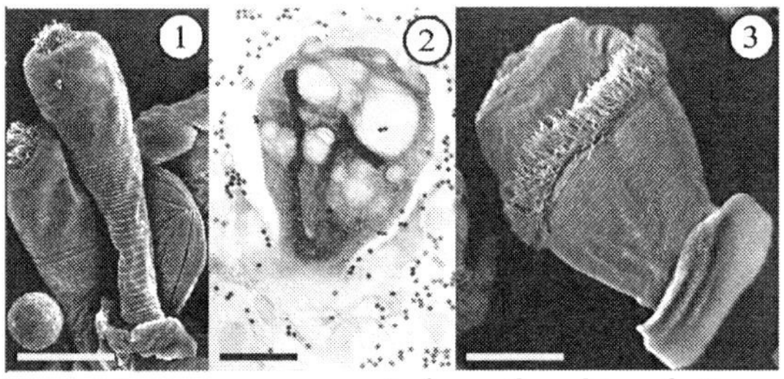

Figure 1. *Apiosoma* sp. A from the skin of *Perca fluviatilis*. Figures 2 and 3. *Ambiphrya* sp. A and sp. B from the gills of *P. fluviatilis*. (1 and 3 - scanning electron micrographs; 2 – light micrograph of haematoxylin-stained specimen). Scale bars = 15 µm.

Corresponding author: bassonl@ufs.ac.za

EFFECTS OF GINGIPAIN R1 AND LPS FROM *Porphyromonas gingivalis* ON THE HUMAN COAGULATION SYSTEM

M. Nunes, T. Fillis, C. Venter and E. Pretorius

Department of Physiological Sciences, Stellenbosch University, Stellenbosch, South Africa

It is well documented that a general hypercoagulable state is present in inflammatory diseases like Parkinson's and Alzheimer's disease (AD), Type 2 diabetes and rheumatoid arthritis[1,2]. In recent studies, Dominy *et al.* established a link between the bacterium *Porphyromonas gingivalis* and AD[3]. *P. gingivalis* has been implicated in periodontitis from which it enters the circulation and is transported throughout the body[4].

P. gingivalis is a gram negative anaerobic bacterium that uses oligopeptides as it main nutrients, obtained via protease activities[4]. Gingipains, a *P. gingivalis* protease, plays an important role in the breakdown of proteins, cell membrane adherence and colonization. The gingipain protease is classified into either arginine-specific (RgpA and RgpB) or lysine-specific (Kgp) groups, with RgpA identified as the more virulent. This protease can enter the circulatory system and interact with various clotting proteins like fibrinogen. Another way *P. gingivalis* can affect the coagulation system is through the release of its inflammagenic lipopolysaccharide (LPS)[4]. LPS is derived from the membranes of the bacterium and can cause amyloidgenic changes in fibrin(ogen)[5]. Thus the aim of this study was to investigate the effects the *P. gingivalis'* protease and LPS had on the coagulation system using scanning electron microscopy (SEM), confocal laser scanning microscopy (CLSM) and thromboelastography® (TEG®).

Purified fibrinogen (Sigma-Aldrich: F3879) and platelet poor plasma (PPP) were used for the SEM, CLSM and TEG® analysis. Human blood was collected from 40 healthy consenting donors (Ethics number: N19/03/043) and centrifuged to obtain PPP. The purified fibrinogen and PPP were exposed to RgpA (final concentration: 500ng/ℓ) and LPS (final concentration: 10ng/ℓ) for 30 minutes. Naïve control samples were used. The samples were processed according to standard SEM processing procedures followed by mounting, carbon coating, and examined with a Zeiss Merlin FEG-SEM. Mann-Whitney-Wilcoxon tests (p-value ≤ 0.05) were done on SEM fibrin fibre thickness measurements.

For the CLSM, controls and LPS exposed purified fibrinogen, conjugated to Alexa Fluor™488 (ThermoFisher: F13191), together with Amytracker™630 (Ebba Biotech: A630-50EX) were viewed with a Zeiss MP880 CLSM in Airyscan mode. To visualize the fluorescent fibrinogen a 488nm laser with band pass (BP) 420-480 and BP 495-550 filters were used. For the Amytracker™630 a 514nm laser and BP 570-620 and long pass (LP) 645 filters were used. PPP controls and LPS exposed samples were also prepared and the autofluorescence of the fibres was used to view the samples[6]. A 488nm laser and a BP 508-570nm filter were used for the green channel autofluorescence, and a 561nm laser and LP 593nm filter were used for the red channel autofluorescence. Viscoelastic analysis was completed after exposing the PPP to RgpA and LPS as described above, whereafter 340µℓ of the PPP was placed in a cup of the TEG® (TEG® 5000 Hemostasis Analyzer) together with 20µℓ of 0.2M calcium chloride to activate the coagulation process[7]. The process was allowed to run until maximal amplitude was reached.

From the SEM fibre analysis, the LPS caused significantly thicker fibrin fibres compared to the controls that formed denser clots. In contrast, with the addition of RgpA, disintegrated clots were observed. The results from the CLSM indicated significant increase in autofluorescence between the control and LPS exposed samples, indicating structurally altered fibrin fibres. This was also visually confirmed with the purified fluorescent fibrinogen exposed to LPS (Figs 1A and B). It was previously shown by Adams *et al.* that RgpA caused a reduction in clot formation[4]. The TEG® analysis indicated that LPS can induce a hypercoagulble state, where as the opposite was seen in the RgpA exposed samples.

In this study the effects of *P. gingivalis'* protease RgpA and LPS on the coagulation system were investigated. This bacterium can possibly contribute to inflammatory diseases after entering the circulation. This study indicated that RgpA and LPS from the bacterium *P. gingivalis*, causes alterations to the coagulation system.

References:
1. Bester, J. and Pretorius, E. (2016) Sci. Rep. 6, 32188.
2. Belstrøm, D. *et al.* (2011) Infect. Immun. 79(4), 1559.
3. Dominy, S.S. *et al.* (2019) Sci. Adv. 5(1), 1.
4. Adams, B. *et al.* (2019) Front Aging Neurosci. 11, 210.
5. Pretorius, E. *et al.* (2017) Cardiovasc. Diabetol. 16(1), 1411.
6. Page, M.J. *et al.* (2019) Sci. Rep. 9(1), 3102.
7. Van Rooy, M.-J. *et al.* (2015) Cardiovasc. Diabetol. 14(86), 1.

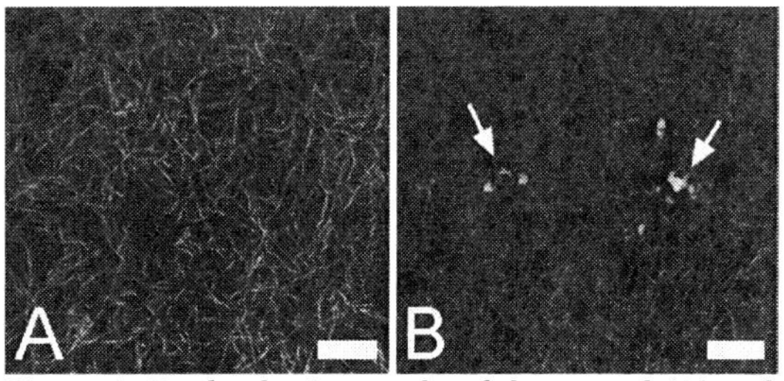

Figure 1. Confocal micrographs of the control (A) and LPS exposed (B) purified fibrinogen. Arrows indicate altered fibrin fibres (Scale bars: 10 µm).

Corresponding author: chantellev@sun.ac.za

EVIDENCE OF CO-LOCALIZATION OF HAEM-DRUG COMPLEX WITHIN A MALARIA PARASITE, USING EFTEM AND RAMAN SPECTROSCOPY.

R. Openshaw[1], M. Waldron[2], M. Jaffer[2] and T. Egan[1]

[1]Department of Chemistry, University of Cape Town, Cape Town, [2]Aaron Klug Centre for Imaging and Analysis, University of Cape Town, Cape Town,

The malaria parasite *Plasmodium falciparum* has a unique pathway for removal of the overabundance of haem from haemoglobin ingested from the hosts' red blood cells. The parasite detoxifies the excess haem by dimerising the haem monomers to form highly insoluble crystals in the form of haemozoin[1]. Haemozoin is central to understanding the biology of malarial disease and ways are being explored to inhibit its formation. Experimental anti-malarial compounds, such as N-(4-(1H-benzo[d]imidazole-2-yl)phenyl)-5-bromonico tin-amide (FAB 5) used in this study, inhibit hemozoin formation.

FAB 5 has been shown to accumulate with exchangeable haem in the parasite digestive vacuole (R. Openshaw unpublished). In this study, we attempt to provide evidence of haem and FAB 5 forming a complex in the vacuole, rather than merely co-localizing. TEM, EFTEM (Energy Filtered TEM) and Raman spectroscopy were utilized to examine the possibility of a haem-drug complex within the digestive vacuole. EFTEM was considered the best technique to map the localization of iron and bromine in the cells.

Parasites were cultured in red blood cells and treated with FAB 5 containing a bromine atom suitable for detection with EFTEM. Samples were subjected to high pressure freezing, cryofixed samples were freeze substituted, followed by embedding in Spurrs resin. Thin sections were collected on copper grids and imaged using a FEI T20 TEM operating at 200kV. The Fe-distribution EFTEM image was collected at an energy offset of 723 eV corresponding to the L –edge using a slit width of 30 eV. The Br-distribution was collected at an energy offset 72 eV corresponding to the M-edge (slit width 10 eV) and/or at 1570 eV corresponding to the L-edge with slit width 40 eV.

Raman spectra and maps were collected on cells deposited onto glass slides. Untreated, parasitized and treated red blood cells were examined and spectra collected with a WiTec confocal Alpha 300 Raman system using a YAG laser at 532 nm.

EFTEM indicated that both iron and bromine co-localized in the parasite digestive vacuole (Fig. 1), indicating that haem or haemozoin and FAB 5 occur in the same place within the vacuole. Raman spectroscopy was employed in order to determine if the iron indicated in the EFTEM was haem or haemozoin. Initially, spectra were collected from pure haemozoin and a synthetically prepared haem-FAB 5 complex (Fig. 2). These data were overlaid on Raman maps collected from the red blood cells in order to indicate where the haemozoin and haem-drug complex occurred within the cell. Figure 3 shows that both in infected and treated red blood cell, haemozoin and haem are present in the centre of the cell where the parasite was situated. This indicates that haem and haemozoin co-localise in the cell vacuole.

The results from the TEM, EFTEM and the Raman spectra strongly suggest that there is a possible drug-haem complex within the malaria parasite.

References:
1. Martiney, J. *et al.* (1996) Mol. Med. 2, 236.
2. L'abbate, F. *et al.* (2018) Eur. J. Med. Chem. 159, 243.

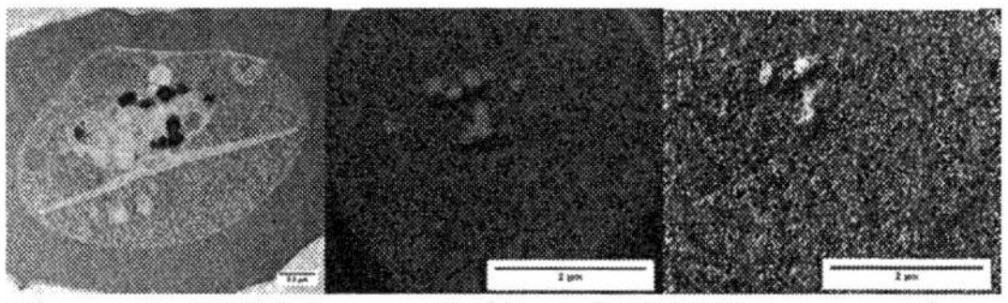

Figure 1. TEM image (left) and EFTEM maps of an infected red blood cell, map of iron (centre) and bromine (right).

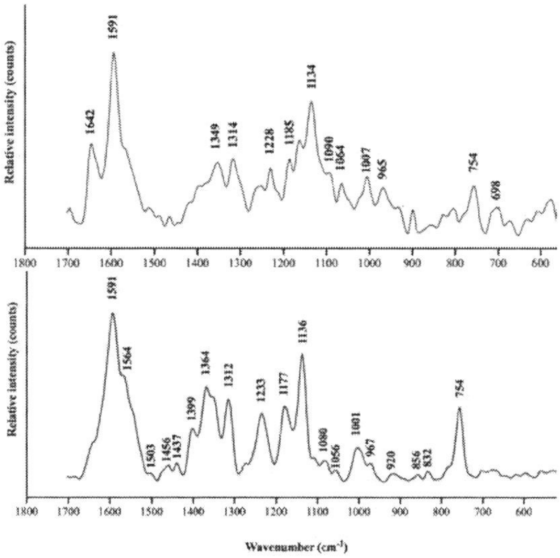

Figure 2. Raman spectra of haemozoin (green) and haem-FAB 5 complex (purple).

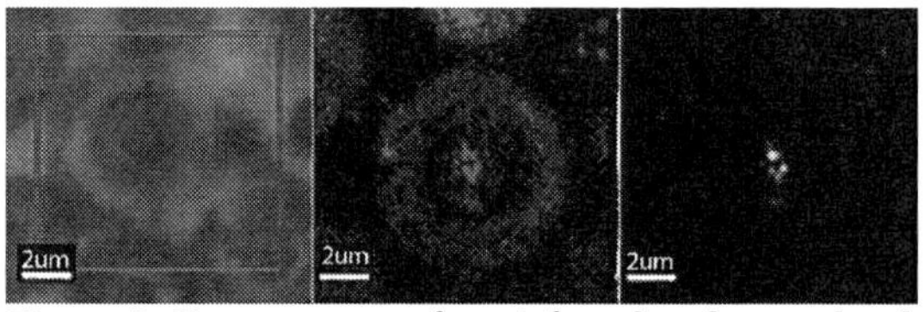

Figure 3. Raman maps of an infected and treated red blood cell (left), haemozoin (centre) and haem-FAB 5 complex (right).

Corresponding author: miranda.waldron@uct.ac.za

FIRST RECORDS OF THE GENUS *Polycycla* (MOBILIDA: URCEOLARIIDAE) FROM HOLOTHURIAN HOSTS IN AFRICA

G.P. de Jager and L. Basson

Department of Zoology and Entomology, University of the Free State, Bloemfontein

Mobilines are ciliated protozoan symbionts that are associated with a wide range of aquatic hosts, invertebrate and vertebrate, freshwater and marine. This order comprises three families, two of which, Trichodinidae and Urceolariidae, have been reported from southern Africa. The genus *Polycycla* (Urceolariidae) is endo symbionts living in the midgut of holothurians (sea cucumbers) and by far the most understudied of the mobilines. This study is the first to successfully stain and measure the morphological characteristics needed for taxonomy and also the first to capture electron photomicrographs of this genus.

Two different holothurian hosts, *Roweia frauenfeldi frauenfeldi* (Ludwig, 1882) and *Pentacta doliolum* (Pallas, 1766) was collected from various localities along the southern coastline of South Africa. The midgut from these hosts were dissected out and the polycyclids were collected via a glass pipette and stored in various fixatives (10% buffered formalin and 4% glutaraldehyde for SEM; Bouin's solution and dry smears for compound microscope morphometrics; absolute ethanol for molecular analysis). Dry smears were impregnated with $AgNO_3$ for the hard denticular structure analysis[1], Bouins fixed smears were stained with Mayer's hæmatoxylin for nuclear material measurements[2], according to Lom's protocol for mobiline taxonomy[3]. Formalin and glutaraldehyde fixed samples were prepared for scanning electron microscopy. All photomicrographs were taken with a Zeiss Axiophot compound microscope and AxioCam ICc 5 camera and a JEOL WINSEM JSM 6400 scanning electron microscope (at 5kV).

The body shape of this mobiline reminds of the ancient Greek *pilos* (πῖλος) type cap; conical and brimless (Figs 1C & 2A). Body surface (pellicle) shows definite ornamentation that seems smooth in relaxed specimens (Fig. 2A). The compound ciliary girdle on the aboral side (Fig. 2A) consists of an inner row of short cilia, a compound ring of multiple rows of varying lengths and a single interspaced ring of marginal cilia. On the opposite side the adoral cilia is produced in a very small spiral where both haplo- and polykineties spiral anticlockwise into the infundibulum (Fig. 2B). The studied polycyclids revealed a denticle ring consisting of several narrow and numerous plate-like structures (Fig. 1A). The nuclear apparatus contains a large prominent macronucleus that is Ω (Omega) to angular C-shaped (Figs 1B & C). The body of the macronucleus appears as round or elongated interconnected beads. A small, but prominent, micronucleus (Fig. 1B) is visible in the $-y^1$ position[3].

Only one species of *Polycycla* is known from literature, i.e. *Polycycla synaptae* (Cuénot, 1891) from a holothurian from France[4], with the most recent record from the mid-1960's[5]. The present study represents the most comprehensive investigation using $AgNO_3$ impregnation and electron microscopy that revealed never before seen morphological details. These include the unique structure and shape of the body, morphological features of the ciliary organelles and the hitherto undiscovered characteristics of the denticle disc. The two holothurians each harboured morphologically distinct mobiline species, both of these new, indicating host specificity.

In conclusion, the results of this study not only revealed unique new features never before observed, but will also triple the number of species from a previously thought monophyletic genus.

References:
1. Klein, B.M. (1958) J. Protozool. 5, 99.
2. Humason, G.L. (1979) Animal Tissue Techniques. San Francisco, W.H. Freeman and Company.
3. Lom, J. (1958) J. Protozool. 5, 251.
4. Cuénot, L. (1891) Rev. Biol. 3, 81.
5. Fauré-Fremiet, E. and Czapik, A. (1966) Cah. Biol. Mar. 7, 61.

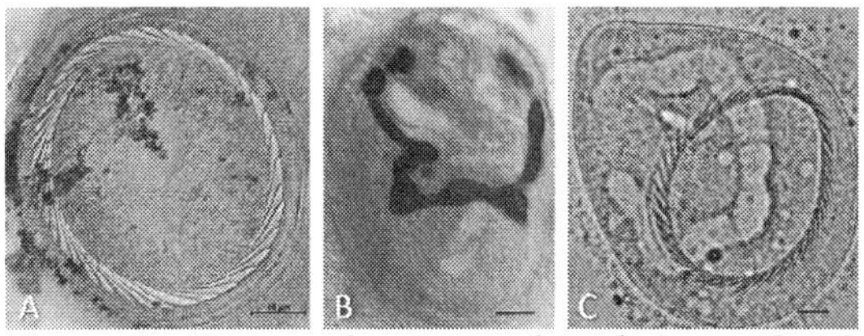

Figure 1. Compound micrographs of A – Aboral denticle ring, B – nuclear apparatus and C - live observation of *Polycycla* sp. A from *Roweia f. frauenfeldi* (scale = 10µm).

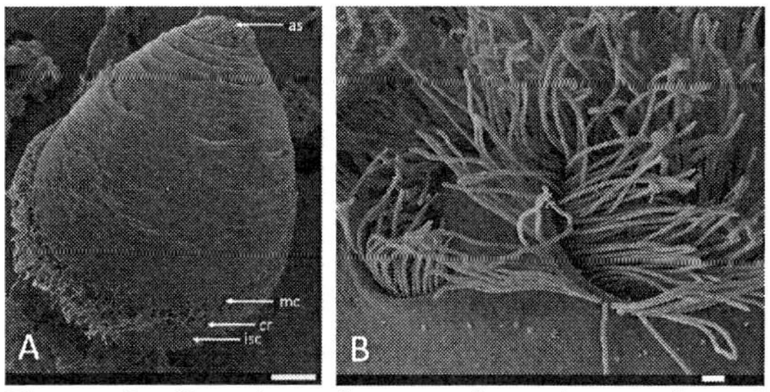

Figure 2. Scanning electron photomicrographs of polycyclids; A - lateral view of *Polycycla* sp. A (host: *Roweia f. frauenfeldi*) with inner row of short cilia (isc), marginal cilia (mc), compound ring (cr) and adoral spiral (as); B – Adoral spiral of *Polycycla* sp. B (host: *Pentacta doliolum*) (scale = 10µm).

Corresponding author: bassonl@ufs.ac.za

FOLIAR MICROMORPHOLOGY, ULTRASTRUCTURE AND HISTOCHEMICAL ANALYSIS OF *Barleria albostellata*

S. Gangaram[1], Y. Naidoo[1] and Y.H. Dewir[2,3]

[1]School of Life Sciences, University of KwaZulu-Natal, Westville, Durban, [2]Plant Production Dept, College of Food and Agricultural Sciences, King Saud University, Riyadh, Saudi Arabia, [3]Dept. of Horticulture, Faculty of Agriculture, Kafrelsheikh University, Egypt

Medicinally-important active compounds are found in morphologically-distinct plant structures known as trichomes[1,2,3]. Trichomes produce, store and/or secrete secondary metabolites mainly to deter herbivores. Trichomes can occur on the plant surface as hairs or external glands. Plant secretory structures aid as taxonomic characters, assisting in the identification of plant families[4].

Barleria albostellata (Acanthaceae), a species indigenous to South Africa, is a valuable medicinal plant with a broad spectrum of antibacterial activity as well as anti-inflammatory properties[5]. This study aimed to characterise the micromorphology, distribution and chemical composition of trichomes present on the leaves and stems of *B. albostellata* using light and electron microscopy and histochemistry. Three stages of leaf development were compared: emergent, young and mature. Each stage was differentiated based on leaf length, colour and texture.

Fresh leaves and stems were examined using a Nikon AZ100 stereomicroscope. The micromorphology of chemically-fixed samples of both leaf surfaces (± 5 mm^2) for each developmental stage and stems were examined using a LEO 1450 scanning electron microscope. Samples were processed conventionally, mounted onto stubs, coated with gold and viewed. Additionally, conventional transmission electron microscopy preparation techniques were used to fix and section leaves and stems for ultrastructural investigations. For histochemistry, fresh leaf and stem sections (90-110 μm) were stained using Wagner and Dittmar reagent (alkaloids), Ferric Trichloride (phenolics), Ruthenium Red (mucilage), Nile Blue (acidic lipids) and Sudan Black (lipids, lignin and suberin). Stained sections were viewed and images captured with a Nikon Eclipse 80i compound light microscope. For fluorescence microscopy, sections were stained with Acridine Orange (to assess cell viability) and viewed using a Nikon DS-Fi1 compound microscope.

Morphological observations using stereomicroscopy and SEM revealed a dense indumentum bearing numerous non-glandular trichomes on the leaves and stems. The density of both glandular and non-glandular trichomes appeared to decrease with progressive leaf development. Histochemically stained sections and SEM micrographs revealed a dense indumentum with the presence of five morphologically distinct glandular capitate trichome types, *viz.*: Type I, a 4-celled broad secretory head with a short stalk embedded in the epidermal layer; Type II, a 2-5-celled secretory head, with a 1-2-celled short stalk; Type III, a bulbous secretory head with a long-stalk; Type IV, a broad secretory head with several cells, a neck cell and a 3-celled long stalk; and Type V, two glandular capitate trichomes were attached to each other and branched. Multangulate-dendritic branched non-glandular trichomes (Fig. 1) and glandular heads attached to branched non-glandular trichomes were also present on the leaves and stems.

Ultrastructural investigations showed well defined organelles such as plastids, mitochondria and endoplasmic reticulum, indicative of metabolically active tissue. Stained leaf and stem sections indicated the presence of major chemical compounds of medicinal importance (alkaloids and phenolics). These compounds may serve as herbivore deterrents. Sections stained with Acridine Orange illustrated trichome viability (ability to metabolize chemical compounds) of the glandular and non-glandular trichomes by fluorescing yellow-green. Our results showed that the presence of both glandular and non-glandular trichomes in *B. albostellata* are important diagnostic features for taxonomists. To our knowledge, this study will represent the first detailed report to describe key micromorphological features of the foliar structures of *B. albostellata* as well as the preliminary chemical composition of the secretions produced by these structures.

References:
1. Gurib-Fakim, A. (2006) Mol. Aspects Med. 27(1), 1.
2. Levin, D.A. (1973) Q. Rev. Biol. 48(1), 3.
3. Thanh, T.L. *et al.* (2017) J. Plant Interact. 12(1), 108.
4. Svoboda, K.P., Svoboda, T.G. and Syred, A.D. (2000) A review and atlas of micrographs. Wales, Microscopix Publications.
5. Amoo, S.O., Finnie, J.F. and Van Staden, J. (2009) J. Ethnopharmacol. 121(2), 274.

Figure 1. Scanning electron micrograph of a multangulate-dendritic branched non-glandular trichome of *B. albostellata*.

Corresponding author: serishagangaram@yahoo.com

IMPACT OF ETHINYLESTRADIOL AND GESTODENE ON VISCOELASTICITY AND ULTRASTRUCTURE OF WHOLE BLOOD

A.C. Swanepoel, H.C. Roberts and J. Bester

Department of Physiology, School of Medicine, Faculty of Health Sciences, University of Pretoria

Combined oral contraceptives (COCs) are used worldwide for both it's contraceptive and non-contraceptive benefits such as protection against endometrial and ovarian cancers as well as the treatment of acne vulgaris, dysmenorrhea and endometriosis-associated pain, to name but a few[1]. Unfortunately, these synthetic hormone formulations are also associated with increased risk of venous thrombosis (VT)[2]. "Red clots", consisting of fibrin and red blood cells (RBCs), are the hallmark of VT[3]. While it is recognised that COC use increases the risk of VT, the specific mechanism responsible is not fully understood.

The aim of this study was to examine the effect that ethinylestradiol (EE) in combination with the third-generation progestin, gestodene (GSD), has on the viscoelastic and biophysical properties of whole blood.

Both females using a specific EE-GSD containing COC (n=10) as well as females using no hormonal contraceptives (n=10) were recruited for this study. Blood samples were taken in citrated blood tubes (0·5 mL of 3·8 % sodium citrate) for laboratory analysis after informed consent was obtained. Exclusion criteria were smoking, thrombotic disease and use of chronic medication (except for the COC). Thromboelastography (TEG) was used to determine the viscoelastic properties while light microscopy (LM) and scanning electron microscopy (SEM) were employed to study the biophysical characteristics of whole blood. TEG was performed with a Thromboelastograph 5000 Hemostasis Analyzer System, using the method described previously[4]. Whole blood smears made on glass microscopy slides underwent standardized LM staining with methylene blue and eosin. Samples were viewed with a Zeiss AXIO Imager.M2 light microscope and axial ratios of RBCs were determined as previously described[4]. Whole blood smears prepared on round glass coverslips were routinely prepared for SEM, mounted on aluminium stubs and carbon coated. Samples were viewed using a Zeiss Crossbeam 540 Field Emission Gun (FEG) Scanning Electron Microscope (SEM) at 1kV.

TEG analysis revealed a hypercoagulable profile for the COC group compared to the control group. Figure 1 show representative LM micrographs of whole blood control and COC smears, respectively. Specialised software labels each RBC, determines the circumference and draws the major and minor axis for each cell to calculate the axial ratio. COC use significantly decreased the roundness (1.18 in the control group and 1.13 in the COC group, p-value = 0.005) as well as the circumference (138.8 in the control group and 132.5 in the COC group, p-value = 0.029) of RBCs. SEM confirmed the changes to the RBC shape along with RBC clumping and spontaneous fibrin formation. Figure 2 shows a SEM micrograph showing a RBC trapped in spontaneously formed fibrin (without the addition of thrombin) representative of the impact of EE-GSD containing COCs on whole blood.

Therefore, EE in combination with GSD not only brings about changes to the RBC morphology, which impairs their critical function in circulation, but also causes spontaneous formation of fibrin which explains the increased formation of "red clots" associated with these COCs. Our findings therefore provide a possible mechanism for the increased risk of VT associated with EE-GSD-formulations.

References:
1. Dayal, M. and Barnhart, K.T. (2001) Semin. Reprod. Med. 19(4), 295.
2. O'Brien, S.H. (2014) Semin. Thromb. Hemost. 40(1), 66.
3. Byrnes J.R. and Wolberg A.S. (2017) Blood 130(16), 1795.
4. Emmerson O. *et al.* (2018) Microsc. Microanal. 24(6), 713.
5. Swanepoel A.C., Emmerson O. and Pretorius E. (2017) Microsc. Microanal. 23(3), 599.

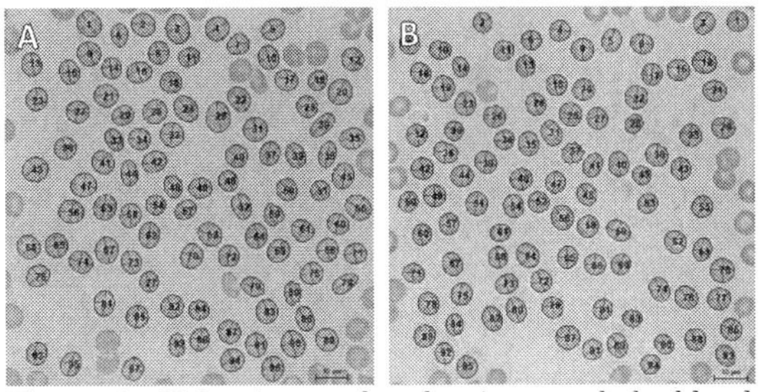

Figure 1. LM micrographs showing a whole blood smears representative of A) the control group and B) the COC group.

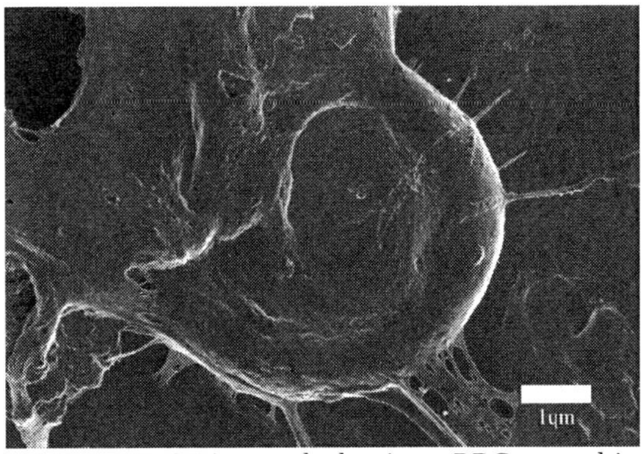

Figure 2. SEM micrograph showing a RBC trapped in spontaneously formed fibrin (without the addition of thrombin) representative of the impact of EE-GSD containing COCs on whole blood.

Corresponding author: albe.swanepoel@up.ac.za

INVESTIGATING THE AUTOPHAGY MACHINERY AND ITS CARGO – A CLEM APPROACH

B. Loos[1], L. Collinson[2], C. Peddie[2], J. Kriel[1], L. Engelbrecht[3], L. Joubert[3], M. Russel[2] and D. Lumkwana[1]

[1]Department of Physiological Sciences, Stellenbosch University, South Africa , [2]Science Technology Platform, Electron Microscopy, Francis Cricks Institute, London, United Kingdom, [3]Central Analytical Facility, Stellenbosch University

Dysfunctional autophagy is associated with a multitude of human diseases, such as protein aggregation in neurodegeneration due to decreased or inhibited autophagy[1], or heightened autophagy in the majority of cancers, also termed 'autophagy addiction'[2]. The rate of protein degradation through the autophagy pathway is termed autophagic flux, where usually the autophagosome machinery, i.e. the abundance of autophagosomes, is considered as an indicator of the degradative activity[3]. Although molecular defects within the autophagy pathway, such as the dysfunction of autophagosome/lysosome fusion, have been investigated in great detail, an accurate assessment of autophagosomes and their cargo within the cellular ultrastructural context has remained limited. Moreover, the accurate identification of autophagic pathway intermediates, in particular autophagosomes and autolysosomes, at electron microscopy level has remained challenging. Given the fact that electron microscopy remains a gold standard in autophagy research, the precise identification of these pathways intermediates is of critical importance. The aim of this project was therefore to employ a correlative light and electron microscopy (CLEM) approach based on structured illumination microscopy (SR-SIM) and serial block-face SEM (SBFSEM) to acquire images and identify key components of the autophagic machinery and their cargo with high specificity and ultrastructural detail.

Mouse embryonic fibroblast (MEF) cells stably expressing GFP-light chain 3 (LC3) were maintained under standard tissue culture conditions, using Dulbecco's Modified Eagles Medium (DMEM), supplemented with 10% foetal bovine serum (FBS) and 1% penicillin-streptomycin (PenStrep). Cells were seeded in 35 mm gridded coverslip dishes (MatTek), incubated with 50 nM lysotracker-blue for 2 h (Molecular probes), followed by a washing step in warm PBS and subsequent gentle fixation using 4% paraformaldehyde in a 1:1 concentration with DMEM, for 10 min at 37 °C. Next, cells were blocked with 5% donkey serum, incubated with acetylated tubulin for 90 min, using 1:100 primary antibody (Santa-Cruz), followed by 30 min secondary antibody incubation (anti-mouse Alexa Fluor 568) as well as anti-phalloidin (Alexa Fluor 647 Phalloidin). Cells were acquired using a LSM 780 P.S.1 station. Briefly, thin (0.1 µm) Z-stacks of high-resolution image frames were collected in 5 rotations using an alpha Plan-Apochromat 100×/1.46 oil DIC M27 ELYRA objective and an Andor EM-CCD camera (iXon DU 885). Images were reconstructed using ZEN Black software (edition 2012) based on a structured illumination algorithm. Next, the gridded coverslips were stained with heavy metals[4], embedded in resin, and after hardening dipped into liquid nitrogen to allow quick removal of the coverslip to expose the monolayer of cells at the block-face. Images of cells were acquired from sequential 15 nm sections using a 3View2XP (Gatan, Pleasanton, CA) attached to a Sigma VP SEM (Zeiss). Image processing, alignment and registration was performed using Fiji and the ec-CLEM plugin utilizing icy open software. Twenty to thirty recognizable cellular landmarks were selected and image overlay transformation was performed.

Our results reveal autolysosomes and lysosomes with distinct electron dense cargo (Fig. 1). Of note, the autolysosome, identified by LC3-positive and lysotracker-blue positive signal (Fig. 1 B and D), is characterized by complex, multilamellar, often circular cargo, of varying electron density. The cargo within lysosomes is characterized by a homogeneous, highly electron dense content (Fig. 1 A and C). These data demonstrate that CLEM enables the identification of autophagy pathway intermediates at electron microscopy level with high precision, informed by the biological functional detail derived through the fluorescence signal. Here we show that, in particular autophagosomes and autolysosomes, can be distinguished clearly through a CLEM approach, allowing future work to characterize specific cargo associated with proteotoxicity and aggregation of disease specific cargo.

References:
1. Hara, T. *et al.* (2006) Nature. 441(7095), 885.
2. Mizushima, N. et al. (2008) Nature. 451(7182), 1069.
3. Du Toit, A. *et al.* (2018) Autophagy. 14(6), 1060.
4. Russell, M.R. *et al.* (2017) J. Cell Sci., 130(1), 278.

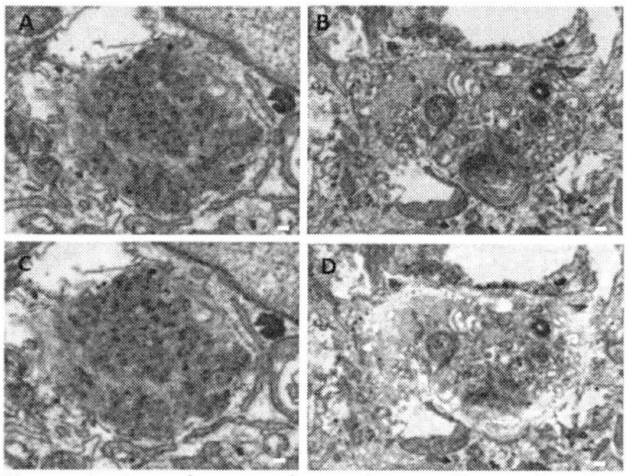

Figure 1. Autophagy pathway intermediates showing electron micrographs (A, B) and correlated SR-SIM and EM micrographs (C, D), indicating a lysosome (A, C) and an autolysosome (B, D). Scale bar: 0,1 µm.

Corresponding author: bloos@sun.ac.za

LATICIFERS IN *Tabernaemontana ventricosa*: MICROMORPHOLOGY, HISTOCHEMICAL ANALYSIS AND ULTRASTRUCTURE

C.M. Naidoo[1], Y. Naidoo[1] and Y.H. Dewir[2,3]

[1]School of Life Sciences, University of KwaZulu Natal, Durban , [2]Plant Production Department, P.O. Box 2460, College of Food & Agriculture Sciences, King Saud University, Riyadh 11451, Saudi Arabia , [3]Department of Horticulture, Faculty of Agriculture, Kafrelsheikh University, Kafr El-Sheikh, 33516, Egypt

Tabernaemontana ventricosa is a medium-sized latex-bearing tree belonging to the family Apocynaceae. Latex is produced in all parts of *T. ventricosa* and functions to deter herbivores, provides resistance to diseases and has potential medicinal value[1]. Latex occurs within specialized secretory cells identified as laticifers [5]. The occurrence of non-articulated laticifers has been noted in the literature for the Apocynaceae, however, the research is limited. This study aims to investigate the morphology, ultrastructure and phytometabolites within the laticifers of *T. ventricosa* from KwaZulu-Natal.

The anatomy of laticifers was determined using fresh leaf sections (1cm^2). Samples were quenched in liquid nitrogen slush (-210°C), fractured into cross- and longitudinal sections and freeze-dried in an Edwards Modulyo freeze-dryer to room temperature over 72 hours. Samples were mounted, gold-coated using a Quorum 150 RES sputter coater and viewed using a Leo1450 scanning electron microscope at a working distance of 14-17mm.

Histochemical analysis and fluorescence microscopy were conducted on fresh leaf material (80-100 µm) using the following tests: Sudan IV (lipids), Ferric Trichoride (phenolics), Ruthenium Red (mucilage), Nile Blue (neutral and acidic lipids), Mercuric Bromophenol Blue (proteins), NADI reagent (essential oils) and Wagner and Dittmar reagent (alkaloids). Stained sections were viewed using a Nikon Eclipse 80i light compound microscope and fluorescence sections were analysed using a Nikon DS-Fi1 compound microscope (Nikon, Japan). All images were captured using NIS-D elements software.

Epoxy resin embedded leaf material was sectioned to 100 nm thickness using a Leica EM UC7 Ultra-Microtome, collected on copper grids and stained conventionally. Images were analysed using a JEOL 1010 Transmission Electron Microscope equipped with iTEM software.

The results revealed laticifers as non-articulated branched and unbranched cells that are composed of a single coenocytic cell, which appears as branched networks at maturity. A great degree of laticifer branching patterns was observed in the vegetative organs (Fig. 1). The continuous branching of laticifers results in the formation of a network comprised of multiple laticifers that expand throughout the plant body[2]. Laticifer cells are often accompanied by meristematic and adjacent cells that develop into a composite system comprised of "Y" or "H" conformations [1].

Ultrastructural investigations showed that the occurrence of laticifers displaying intrusive growth is frequently associated with apical tip projections and tiny lateral protuberances that penetrate the middle lamella of adjacent cells[1]. Additionally, the existence of a thickened middle lamella found in laticifers is an indication of intrusive growth[1]. Laticifers are present in all ground and vascular tissues and are closely associated with the phloem of the leaf. It has been noted that laticifers in the leaves usually branch throughout the mesophyll and proceed in the direction of the veins, which often develop towards the epidermis[1,2].

Intense staining of major compound groups such as alkaloids, phenolics, acidic lipids and essential oils were detected in the laticifers and latex (protoplast).These results are consistent with previous studies, as these compounds were previously detected within species belonging to the genus *Tabernaemontana*. These chemical compounds may be lethal or at least discouraging to herbivores and pathogens, and most likely hindering the development of micro-organisms [2].

References:
1. Canaveze, Y. and Machado, S. R. (2016). Int. J. of Plant Sci. 177 (5), 458.
2. Demarco, D. (2015). Acta Biol. Columb. 20(1), 57.
3. Hagel, J. M., Yeung, E. C. and Facchini, P. J. (2008) Trends in Plant Sci. 13(12), 631.
4. Mahlberg, P. G. (1993). The Bot. Rev. 59(1) 1.
5. Pickard, W. F. (2008). New Phytol. 177 (4), 877.

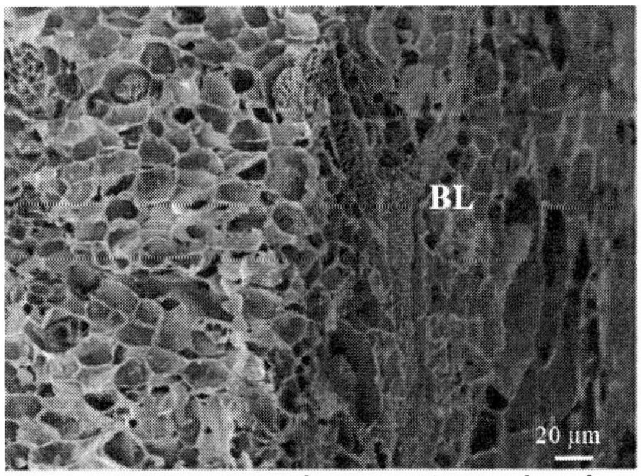

Figure 1. Scanning electron micrograph of a freeze-fractured leaf showing a Y-shaped branched laticifer. Abbreviations: BL= Branched laticifer

Corresponding author: naidooclarissa5@gmail.com

MICROMORPHOLOGICAL, ULTRASTRUCTURAL AND HISTOCHEMICAL PROPERTIES OF FOLIAR TRICHOMES OF *Commelina benghalensis* L.

K. Doolabh[1], Y. Naidoo[1] and Y.H. Dewir[2,3]

[1]School of Life Sciences, University of KwaZulu-Natal, Westville, Durban, [2]Plant Prod Dept, King Saud University, Riyadh, Saudi Arabia, [3]Dept. of Horticulture, Kafrelsheikh University, Egypt

The exploitation of natural products from plants stems from the presence of secondary metabolites which are produced and stored in epidermal appendages such as trichomes[1]. *Commelina benghalensis* L. is a perennial herb found in Africa, and plant parts and/or extracts are commonly used by traditional healers in treating infertility in women[2] and as an anti-inflammatory[2]. Research on the foliar micromorphology of this species is limited. The aim of this study was to examine the micromorphology and histochemical properties of trichomes found on the leaves and stems of *C. benghalensis* L. at three developmental stages, i.e., emergent, young and mature.

Stereomicroscopic analysis was conducted on fresh leaves and stems using a Nikon AZ100 stereomicroscope. Trichome distribution and density were examined. Histochemical analyses were conducted on young leaves sectioned using an Oxford® Vibratome (150-200 µm). The sections were stained with Ferric Trichloride for detection of phenolic compounds and Acridine Orange to detect cell viability. Alkaloids were revealed using the combinatory Wagner's and Dittmar stain. Ruthenium Red tested for mucilage and pectin, Nile Blue for lipids and Calcofluor White for the presence of carbohydrates. Sections were viewed using a Nikon Eclipse 80i compound light equipped with a Nikon DS-Fil camera.

For scanning electron microscopy, leaf segments ($4mm^2$) were plunged into subcooled liquid nitrogen and freeze-dried to room temperature over 72 hours. The segments were viewed using a LEO 1450 SEM. Young leaves were prepared for transmission electron microscopy using standard procedures, sectioned (100nm), stained and viewed using a JEOL 2100 High-Resolution TEM.

Stereomicroscopy revealed red hairs at the apex of the leaf sheath. Entire leaf margins contained only non-glandular trichomes. During leaf maturation, the density of trichomes decreased. It is suggested that emergent leaves need increased protection as they are more susceptible to pathogen and insect attack[3]. The stem had a trichome density similar to the emergent leaves as both were higher compared to the young and mature leaves.

Histochemical tests conducted on stained sections showed important medicinal compounds, *viz.* phenols and alkaloids (Fig. 1). These compounds were predominantly localized in the basal cells of trichomes. SEM revealed non-glandular trichomes that were bi-cellular, multi-cellular and hooked (Fig. 2). The spine-like and hooked nature of non-glandular trichomes could allow for direct impaling of an insect's body which could hinder insect feeding behavior[4].

Transmission electron microscopy of trichomes showed organelles such as well-extensive endoplasmic reticulum profiles and chloroplasts containing plastoglobuli and starch grains, suggesting high metabilic activity.

There is a dearth of scientific investigations on the foliar structure and ultrastructure of *C. benghalensis* L. by South African researchers. Thus, our research presents part of a broader aim in evaluating the micromorphology and medicinal use of this species.

References:
1. Svoboda, K.P., Svoboda, T.G. and Syred, A.D. (2000) Secretory Structures of Aromatic and Medicinal Plants: A Review and Atlas of Micrographs. Knighton, Microscopix Publications.
2. Ibrahim, J. *et al*. (2010) Ethnobot. Leaflets 14,10.
3. Munien, P., Naidoo, Y. and Naidoo, G. (2015) Planta 242(5), 1107.
4. Levin, D.A. (1973) Q. Rev. Biol 48(1), 3.

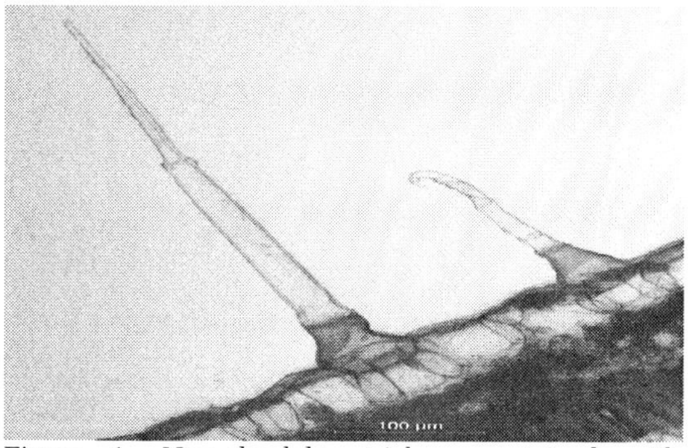

Figure 1. Non-glandular trichomes stained with Ruthenium Red showing mucilage and pectin at the trichome base.

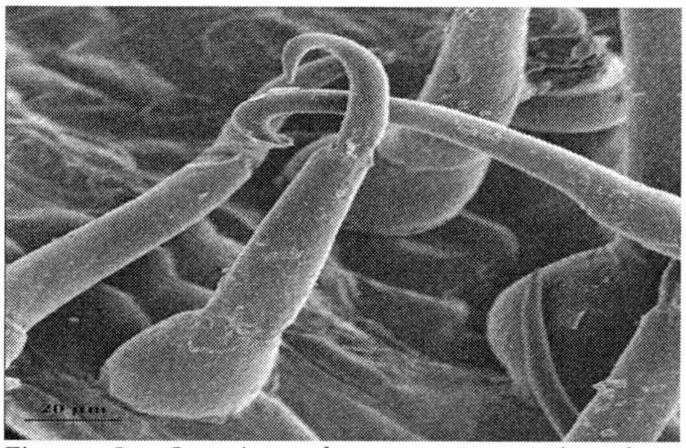

Figure 2. Scanning electron micrograph of a non-glandular, hooked trichome found on *C. benghalensis* L.

Corresponding author: kareshma.doolabh@yahoo.com

MORPHOLOGY OF THE LUNG OF THE RED-EARED TERRAPIN (*Trachemys scripta elegans*)

S. Schoeman[1] and D. Reddy[2]

[1]School of Computer Science and Applied Mathematics, University of the Witwatersrand, Johannesburg, [2]Microscopy and Microanalysis Unit, University of the Witwatersrand, Johannesburg

The red-eared terrapin is indigenous to North America[1]. They are semi aquatic reptiles, which has been classified as an alien invasive species in Southern Africa (under category 1B)[1]. Reptilian lungs display great complexity and diversity[2], with multicameral lungs being the most complex [3]. The red-eared terrapin possess a multicameral lungs which exhibits great amounts of heterogeneity[3].

The present project investigated the morphology of the lung of the red eared terrapin in order to understand how they exploit hypoxic conditions.

The lungs were dissected from the peritoneal cavity of the animals and were sampled using Systematic Random Sampling[4]. The tissue samples were processed using standard histological tecniques[5]. Thereafter sections were cut on a microtome at 5µm and stained using the New Pentachrome staining technique[6]. Post this, they were imaged using bright-field light microscopy.

The lungs of the red eared terrapin are large organs spreading from the top of the carapace to approximately three quarters caudally. The trachea bifurcates into two bronchi that enter the lungs in the cranial ventral region of the lung. The lungs are inverted with the base found cranially while the apex is situated in the caudal region with Intercostal sulci being present on its dorsal lateral surface. The ventral surface is flattened while the dorsal surface is convex where it fits into the shell.

The lungs have a honeycomb[3] appearance with distinct air chambers found around the middle and caudal regions. The smaller caudal region contained one large air chamber while the broader cranial and middle regions are optimized for gas exchange. Small air chambers were found on the periphery of the lung in conjunction to the intercostal sulci.

The larger surface area in the cranial region together with the aforementioned characteristics allows for effective gas exchange in the cranial and middle regions, conduction in the middle region and, storage in the large air chambers of the caudal region. A striated muscular sheath (Fig 1), consisting of several layers, were found on the periphery of the pleura in the cranial region. This coincides with earlier experimental work done on *Chelonia*[7,8]. This sheath was named the *muscularis striatum pulmonale,* which was shown to have a ventilatory function[7,8].

The large air chambers of the caudal region contained smooth muscle septa on the inner surface of the lung pleura. Cartilage was found in the middle and caudal region and ciliated pseudostratified epithelium, with mucin cells interspersed, present in all regions of the lung. Airways were found to be reinforced by smooth muscle knobs and plates with the double capillary system of the blood-gas barrier containing a connective tissue core forming septa throughout the lung. This ensures that the lung remains patent whilst optimal gas exchange occurs.

The smooth muscle in the caudal region will allow for the shunting of air while the striated muscular sheath (MSP) will give the animal the ability to control the pressure in the thoracic cavity[7,8]. Also, together with the smooth muscle airways, the MSP could control the airflow within the lung. This would optimise the overall ability of the terrapin to extract oxygen and store air in a single breath. In addition, by being able to remain in water for longer periods of time, they can maximise the use of the habitat.

References:

1. http://www.invasives.org.za/component/k2/item/572-red-eared-slider-trachemys-scripta
2. Patt, D. and Patt, G. (1969). Comparative vertebrate histology, Harper and Row.
3. Duncker HR. (1978) Proc. in Life Sciences. 1.
4. Reddy, D. (2013). Morphogenetic documentation of the architecture of the Nile Crocodilian lung., Phd thesis, University of the Witwatersrand, Johannesburg, South Africa.
5. Suvarna, S. K., Layton, C., and Bancroft, J. D. (2013). Bancroft's theory and practice of histological techniques. Oxford: Churchill Livingstone Elsevier.
6. Doello K. (2014) Yale J. Biol Med, 87(3), 341.
7. George, J. C. and Shaw, R. V. (1954). J. Anim. Morph. Physiol. 1,13.
8. Shaw, R. V. (1962). A comparative study of the respiratory muscles in Chelonia. Brevoria 161, 1.

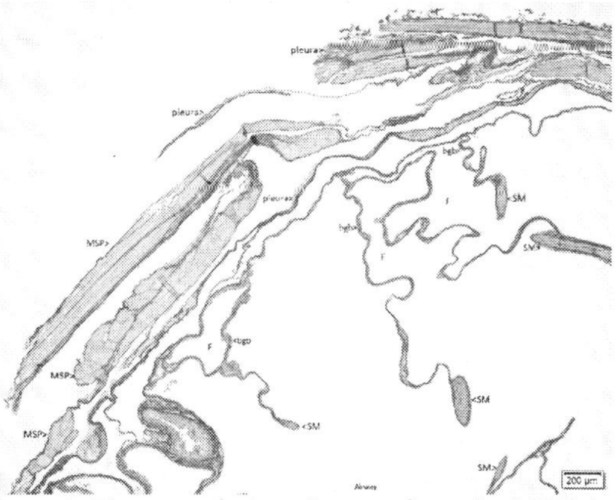

Figure 1. The *muscularis striatum pulmonale* (MSP) is found on the periphery of the lung in the cranial region. The interior surface of the lung contained faveoli (F) with the blood-gas- barrier (bgb) and smooth muscle (SM) knobs and plates forming airways.

Corresponding author: 863372@students.wits.ac.za

NANOPLASTICS IN ZEBRAFISH USING RAMAN IMAGING AND SCANNING ELECTRON (RISE) MICROSCOPY

F. Hipper[1] and M. Waldron[2]

[1]Department of Biology, University of Ulm, [2]Aaron Klug Centre for Imaging and Analysis, University of Cape Town

Plastic pollution in marine systems is ubiquitous and due to its buoyancy and durability is globally distributed across all oceans. However, data on microplastics in freshwater systems is scarce[1]. Many plastics degrade into numerous micro particles between 0.1 and 5,000μm in size. These microplastics share a similar size range with many plankton species and are thus easily ingested by invertebrates and fish. This raises concerns about ecological and health impacts of microplastics across food chains. There is an urgent need to investigate the uptake and distribution of microplastics into marine and freshwater animals.

Microplastic particles are most commonly identified using micro-Fourier transform infrared spectroscopy or near-infrared spectroscopy[2]. Identification of fluorescent polystyrene microbeads in the tissue of Tilapia has been successfully accomplished using a spectrophotometer[3] and Raman spectroscopy has been used to identify plastic strands in crustacean stomachs[4]. However, the identification of plastics in the tissue of fish is challenging, so this preliminary study was initiated in order to discover if polystyrene beads in fish gut and brains could be detected using Raman spectroscopy.

In this study, zebrafish (Cyprinidae) approximately 8 months old were cultivated in four tanks. Polystyrene nanobeads of three sizes (30nm, 50nm and 100nm) were introduced into three of the tanks (tank four was a control tank) and two fish were sacrificed each week for 6 weeks. The fish were frozen in OCT embedding medium, sectioned and the sections placed on glass slides to dry. The OCT was then removed by rinsing the slides gently with PBS buffer and distilled water. The slides were then viewed in the RISE (Raman imaging and scanning electron) microscope.

The samples were viewed in a Tescan MIRA3 SEM under low vacuum conditions, using a backscatter detector. Both the gut and the brain areas of the fish were examined. Once images had been taken in the SEM, the sample was moved from the SEM objective to the Raman objective inside the SEM chamber. A Raman map was acquired of the same area using a WiTec Alpha 300 Raman system. Figure 1 shows Raman maps superimposed on the SEM images, the areas containing polystyrene are shown in red. Figure 2 shows a Raman map of polystyrene superimposed onto a SEM image of the brain of a control fish sacrificed after 6 weeks, showing a small amount of polystyrene present in the control sample.

The results presented in Figure 1 are from fish that had been sacrificed one week after the tank was dosed with 30nm polystyrene beads. Although there appeared to be a small amount of polystyrene detected in the control fish, Figure 1 indicates that polystyrene beads were present in larger concentrations in fish living in water containing the polystyrene beads. Altthough the beads tended to clump in the brain, they were more evenly dispersed in the fish gut. All the other samples from the time series showed the presence of polystyrene in the tissue.

This preliminary study indicates that zebrafish living in water containing polystyrene nanobeads show the presence of beads in tissues within one week. Now that the technique has been perfected in a controlled experiment, the next stage will be to apply the technique to wild caught fish.

The study indicates how RISE can give vital information about the presence of nanoplastics in fish organs.

References:
1. Wagner, M. *et al.* (2014) Env. Sci. Europe 26,12.
2. Hidalgo-Ruz, V. *et al.* (2012) Env. Sci. and Technol. 46(6), 30603.
3. Ding, J. *et al.* (2018) Env. Pol. 238,14.
4. Murray, F. and Cowie, P. (2011) Mar. Pol. Bull. 62(6), 1207.

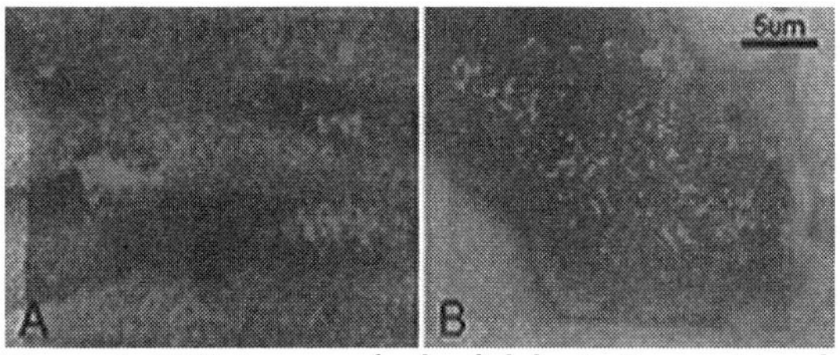

Figure 1. RISE images of zebrafish brain (image A) and gut (image B) showing the Raman map superimposed on SEM images. (Scale bar applies to both images.)

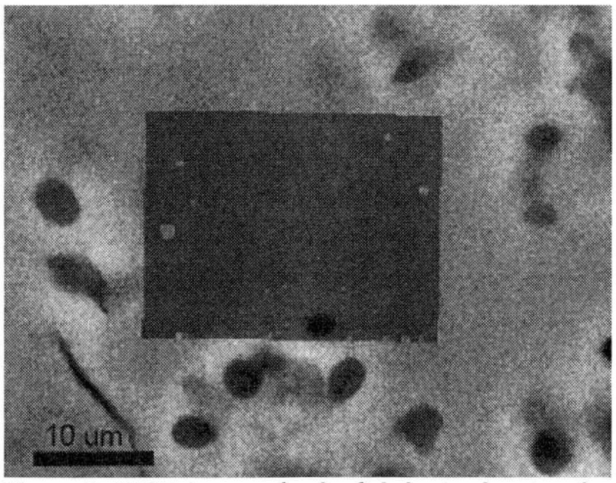

Figure 2. RISE image of zebrafish brain showing the occurrence of polystyrene in the control sample.

Corresponding author: miranda.waldron@uct.ac.za

NEMATODES COLLECTED FROM LEAF LITTER IN JONKERSHOEK NATURE RESERVE, STELLENBOSCH

C. Jansen van Rensburg, G.P. de Jager and L. Basson

Department of Zoology & Entomology, University of the Free State, Bloemfontein, South Africa

The leaf litter layer in forests is full of life, and is known to provide food and shelter to many organisms[1]. These organisms in turn decompose the litter into soluble chemicals and minerals such as nitrogen, calcium and sulphur. The majority of soil nematodes (roundworms) are beneficial and play an important role in the soil food web by controlling nutrient and mineral cycling and regulating the behaviour of microbial communities[2,3]. These soil nematode communities are sensitive to changes in the environment[4]. When studying plant-parasitic and free-living nematodes, nematode ecological and taxonomic investigations have particularly neglected leaf litter as a substrate, focusing primarily on plants and soil respectively.

With this in mind, a nematode survey was undertaken in the Jonkershoek Nature Reserve in December 2018 to firstly map the biodiversity of nematodes occurring within the leaf litter, and secondly to assess whether the nematode community could be used as bio-indicators for selected areas within the reserve.

Approximately 500ml leaf litter samples were collected from seven different localities within the mountainous area of Jonkershoek Nature Reserve. Samples were placed in Jiffy bags, sealed and labelled and kept cool. In the laboratory in Bloemfontein, nematodes were extracted from leaf litter samples by using the modified Baermann tray technique, afterwhich they were processed using a centrifugal sugar flotation method. Extracted nematodes were mounted on Cobb slides for identification using light microscopy. Additional material was prepared for scanning electron microscopy by dehydrating specimens through a series of increasing ethanol concentrations, critical point dried and coated with gold. Nematodes were viewed using a JEOL JSM-7800F Field Emission Scanning Electron Microscope at 5kv.

Currently nematodes are still being identified, however, two nematode genera, *Prionchulus* and *Plectus* belonging to the families Mononchidae and Plectidae respectively, seem to be the dominant genera collected from all sampling localities in the reserve.

The genus *Prionchulus* represents a group of predaceous nematodes, with only one species having been described from this genus in South Africa[5]. *Prionchulus* species collected during this study can be characterised by the lip region being round, with protruding labial and cephalic papillae (Fig. 1a). The dorsal tooth is medium sized (Fig. 1b) and the subventral plate is armed with denticles (Fig. 1b). The amphid is round to oblong (Fig. 1a). The tail is conical and ventrally bent without a spinneret present (Fig. 1c).

The genus *Plectus* can be regarded as a well defined natural group comprised of some of the most common and important soil and aquatic species[5]. The *Plectus* sp. collected during the present study can be recognised by some of the following characters: The labial region is narrow and continuous with the body. The cephalic setae are almost as long as half the labial diameter (Fig. 2a), with the amphidial fovea being round (Figs 2a, b) and situated at mid-stoma. The tail is five to six anal body diameters long, is cylindrical and ventrally curved in the posterior half with three caudal setae on each side. A spinneret is present at the tip of the tail (Fig. 2c).

This is the first report of these two genera from leaf litter in South Africa. Other nematode families that have also been identified include Dorylaimidae, Aphelenchidae, Qudsianematidae and Aporcelaimidae.

References:
1. Johnson, A. and Catley, K.M. (2002) Life in the leaf litter. New York, Center for Biodiversity and Conservation.
2. Bongers, T. and Ferris, H. (1999) Trends Ecol. Evol. 14(6), 224.
3. Ferris, H. (2010) J. Nematol. 42(1), 63.
4. Cao, Z.P. *et al.* (2004) Pedosphere 2(1), 26.
5. Heyns, J. (1971) A guide to the plant and soil nematodes of South Africa. Cape Town, A.A. Balkema.

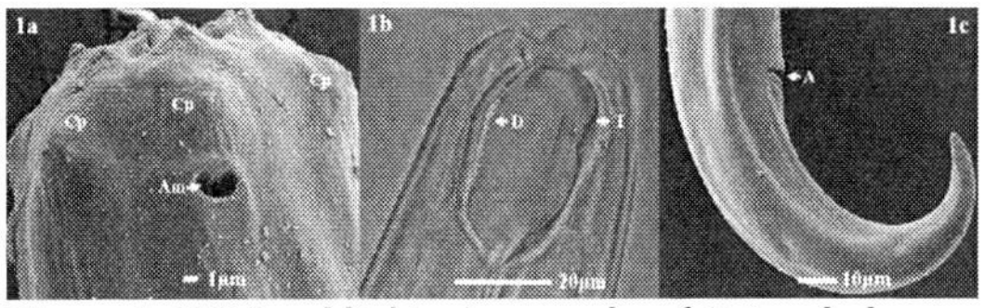

Figure 1. SEM and light micrographs of *Prionchulus* sp. head region (1a, 1b) illustrating positon of cephalic papillae (Cp), amphid (Am), denticles (D), dorsal tooth (T) and tail with anus (A).

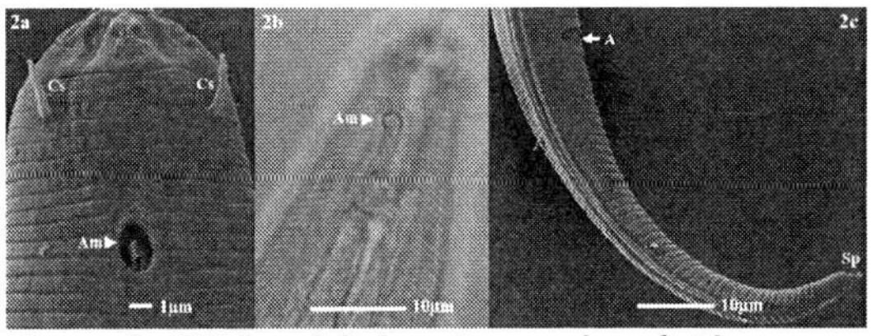

Figure 2. SEM and light micrographs of *Plectus* sp. head region (2a, 2b) illustrating position of cephalic setae (Cs), amphid (Am), and tail (2c) with anus (A) and ending in spinneret (Sp).

Corresponding author: JvRensC@ufs.ac.za

SUCCESSIVE CHARACTERIZATION OF AUTOPHAGIC FLUX IN *Mycobacterium tuberculosis*-INFECTED MACROPHAGES

N. Okugbeni[1], A. du Toit[2], G. Johnson[1], B. Loos[2] and C. Kinnear[1]

[1]Dept. of Biomed. Sci., Stellenbosch University , [2]Dept. of Physiol. Sci, Stellenbosch University

Tuberculosis (TB), is a respiratory disease that caused approximately 10 million new cases in 2017. It is the leading cause of death by a single infectious agent- *Mycobacterium tuberculosis* (*M. tb*)[1]. In recent years, a cellular pathway- Autophagy has been shown to be involved in the killing of *M.tb*. Autophagy is a degradative process in cells that digests and recycles damaged proteins, organelles and invading pathogens[2].

When encountering *M. tb*, host macrophages engulf the invading bacteria for degradation. *M.tb*, however, has developed mechanisms to damage the phagosome, which allows it to escape degradation. The damaged phagosome and escaping bacteria are then targeted for autophagic degradation via the lysosome. *M.tb* has been reported to also escape autophagic clearance[3]. Mechanisms underlying this evasion are still unclear.

To describe this occurrence visually, we characterized autophagic flux in *M. tb*-infected macrophages using immunofluorescence imaging. Cell line macrophages (RAW264.7) were infected with mCherry-expressing *M.tb* and fixed at 4, 24, 48 and 72 hours post infection for antibody staining of 4 different autophagy membrane proteins. A subset of samples were treated with Bafilomycin- a lysosomal inhibitor for comparison. Cells were visualized using a Zeiss LSM 780 confocal microscope. Microscopy was also compared with Western blot and bacterial colony-forming unit (CFU) data.

The imaging data showed that macrophages exhibited heightened autophagic containment of bacteria and flux of proteins in response to infection at 4 hours (Fig. 1). *M. tb* free of autophagic containment was also observed in some cells at this timepoint (Fig. 2). The flux of proteins, however, decreased at 24 hours post-infection. Interestingly, with this decrease in autophagy proteins flux, bacteria autophagic flux continued to increase with the highest flux at 48 hours post infection. At 72 hours, macrophages displayed a varied response with more observable signs of macrophage cell death. Western blot and CFU data supported these observations.

These data reveal the successive flux of both bacteria and autophagic proteins in infected macrophages. This is an improvement on autophagy experiments that study one timepoint only. Colocalization results of *M. tb* with the different autophagy membrane proteins reveal the distinct stages at which *M. tb* evades autophagy degradation. This comprehensive progression of how and when *M. tb* inhibits autophagy flux will enable TB researchers design more accurate and descriptive experiments, providing new information to drug development researchers.

References:
1. WHO (2018) Internet Available from http://www.who.int/mediacentre/factsheets/fs104/en/
2. Gutierrez, M.G. *et al.* (2004) Cell 119(6), 753.
3. Deretic, V. *et al.* (2006) Cell. Microb. 8(5), 719.

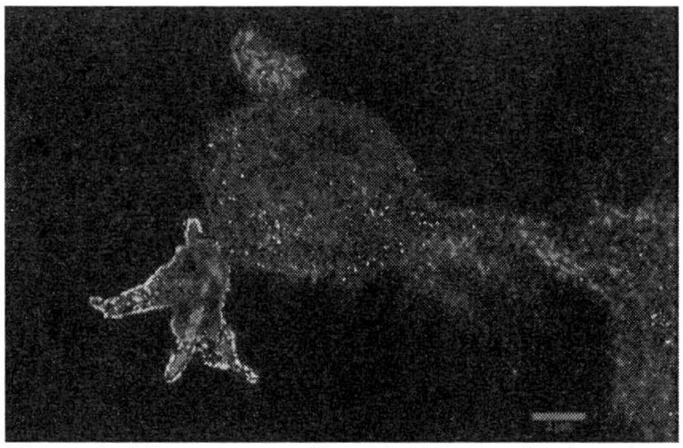

Figure 1. Macrophage cell, with DAPI stained nucleus (blue), Beta actin filaments in purple and LC3 autophagy protein in green, is observed to be encapsulating a group of *M. tb* cells in red. Scale bar = 2 µm

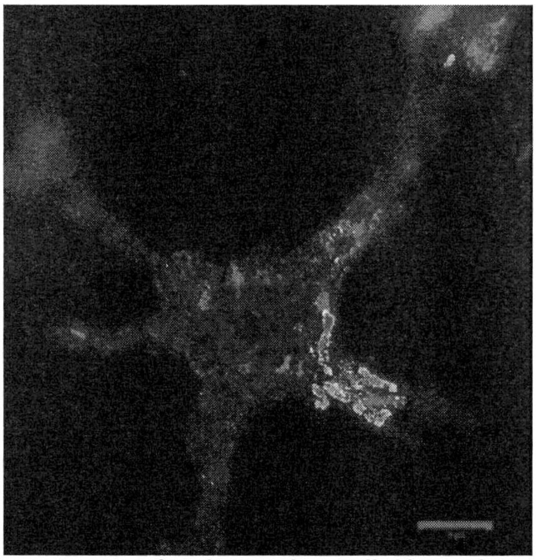

Figure 2. An infected macrophage cell (blue nucleus), Beta actin filaments in purple and LC3 autophagy protein in green. *M. tb* cells (red) are seen to be devoid of LC3 encapsulation, despite the heightened recruitment of LC3 towards some of the bacteria. Scale bar = 5 µm

Corresponding author: gkin@sun.ac.za

THE EFFECT OF MODIFIED CITRUS PECTIN ALGINATE PROBIOTIC ON FAECAL LACTOBACILLI IN BALB/C MICE

F.O. Odun-Ayo[1], J. Mellem[1] and L. Reddy[2]

[1]Department of Biotechnology and Food Technology, Durban University of Technology, KwaZulu-Natal, [2]Faculty of Applied Science, Cape Peninsula University of Technology, Cape Town

Intestinal bacteria utilize the constituent of ingested food for their metabolic activities making the gut microbiota a target in the modification of functional supplement foods. Probiotics have been defined as live microorganisms that, when administered in adequate amounts, confer a health benefit on the host. It was further described as a live microbial feed and food supplement that beneficially affect the host's intestinal tract[1]. Modified citrus pectin (MCP), a natural polysaccharide treated enzymatically to break down galactans into smaller fragments, is used as supplements for cell proliferation. However, studies have shown MCP may have need of an additional food supplements to achieve maximum bioactivity[2]. This study examined the effect of microencapsulated probiotic *Lactobacillus acidophilus* ATCC 4356 with/without MCP and alginate calcium (AC) on the faecal lactobacilli microbiota population in healthy Balb/c mice.

Fresh cell suspensions of *L. acidophilus* ATCC 4356 were prepared for each microencapsulation procedure and enumerated by plating on De Mann, Rogosa and Sharpe (MRS) agar. Modified citrus pectin alginate (MCPA) and AC probiotic microbeads (Fig 1) were produced aseptically and separately at room temperature using emulsification polymerization . Microbeads were broken down by shaking in sterile phosphate buffer solution for 10 min. Cell suspensions in serial dilution (10^1-10^9) were plated and viable number of cell growth was enumerated using an automated colony counter Doc-It imaging station. The microbeads (MCPA and AC) released between 8-9 $\log_{10}$cfu/g of viable *L. acidophilus* ATCC 4356. After ethical approval, seven-week-old male Balb/c mice (n=40) weighing 20-25 g were separated into four groups (n=10). MCPA and AC probiotic, MCP solution (3 g/ml) or water (control) were administered to mice by gavage (intragastric) once daily at a volume of 0.2 ml for 28 days. Faecal lactobacilli counts were determined as described. Data were subjected to ANOVA and Tukey's test, as appropriate, for comparative analysis among the groups using Graphpad Prism software.

The number of faecal lactobacilli in the MCPA probiotic-treated mice significantly increased from day 0 to 14 [7.60 ± 0.23 to 8.32 ± 0.08 ($\log_{10}$cfu/g); $p < 0.001$]. At day 28, a steady increase was noted in the number of faecal lactobacilli in the MCPA probiotic-treated group [8.36 ± 0.23 ($\log_{10}$cfu/g)] while a little or no increase was noted in the AC probiotic- and MCP-treated mice, [8.22 ± 0.02 and 8.08 ± 0.14 ($\log_{10}$cfu/g) respectively ($p > 0.05$), Fig 2]. Limitation to this study is *in vivo* sampling of the colon tissue as some lactobacilli may colonized discrete areas or remain adherent to the mucosal surfaces and therefore remain non-detectable using microbiological agar. The highest number of viable faecal lactobacilli during the treatment period was detected in the MCPA probiotic-treated mice compared to the AC probiotic-, MCP-treated and control groups. This may be due to pectin influence on metabolism in the gut through colonic fermentation and indirect effects of other colonic bacteria[3]. Thus, it is plausible to presume that the disintegration of MCPA by both *L. acidophilus* ATCC 4356 and endogenous microbiota might have induced energy and nutrients for their growth and survival. Modified citrus pectin alginate combined with live probiotic, *Lactobacillus acidophilus* ATCC 4356 supplements can be used as potential probiotic therapy which improves MCP bioactivity and the intestinal lactobacilli microbiota in healthy host.

References:
1. Vasiljevic, T. and Shah, N. (2008) J. Dairy Sci. 18, 714
2. Maxwell, E. G. *et al.* (2012) Trends Food Sci. Tech. 24(2), 64.
3. Wicker, L. *et al.* (2014) Food Hydrocoll. 42, 251.

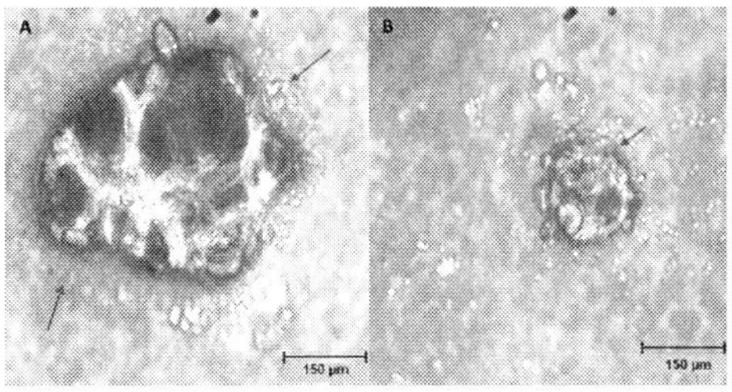

Figure 1. The morphology of (A) modified citrus pectin alginate (MCPA) and (B) alginate calcium (AC) probiotic microbead particles produced by emulsification with arrows indicating the chitosan coating.

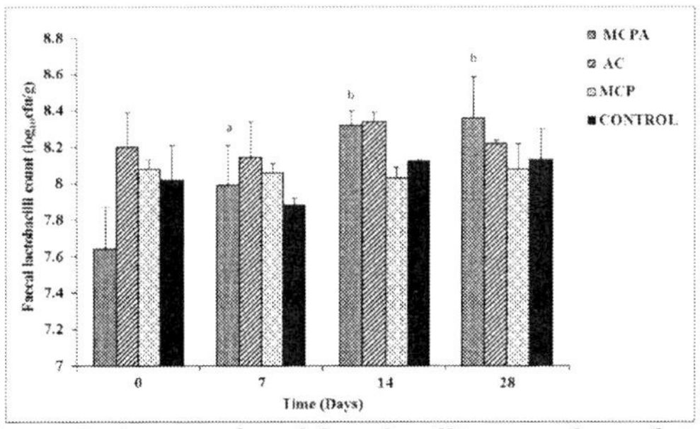

Figure 2. Average faecal lactobacilli count ($\log_{10}$cfu/g) in mice administered with modified citrus pectin alginate (MCPA), alginate calcium (AC) probiotics, modified citrus pectin solution (MCP) and water (control) for 28 days.

Corresponding author: fredodunayo@gmail.com

THE SECRETORY PELTATE GLANDULAR TRICHOMES OF Combretum erythrophyllum (COMBRETACEAE): MICROMORPHOLOGY, ULTRASTRUCTURE AND HISTOCHEMISTRY

S. Bantho[1], Y. Naidoo[1] and Y.H. Dewir[2,3]

[1]University of KwaZulu Natal, [2]King Saud University, [3]Kafrelsheikh University

Combretaceae is a large angiosperm family comprised of approximately 20 genera and 600 species worldwide[1]. The family is prominently found in Asia, Australia and Africa as members thrive in both tropical and subtropical regions[2].

Combretum is the largest and most reported genus of this familydue to the plethora of traditional medicinal uses they offer[3,4]. *Combretum erythrophyllum* (Burch.) Sond, commonly known as the river bushwillow or umdubu (Zulu), is one such species that has been reported to contain secondary metabolites of medicinal importance[5].

Medicinal flora are known to contain secondary metabolites which may effectively improve an individual's state of health. These medicinal properties are attributed to secondary metabolites, which in some instances are exuded by specialized micro-secretory structures such as trichomes, laticifers or glands[6,7]. Screening plants for their potential medicinal properties in relation to the presence of micro-secretory structures is of utmost importance. Thus, this study investigated the micromorphology, ultrastructure and histochemistry of the secretory structures found on the surface of leaves (emergent, young and mature) and stems of *Combretum erythrophyllum*.

Stereo-, light, scanning electron and transmission electron microscopy techniques were used to evaluate the micromorphology, ultrastructure and histochemistry of secretory structures. The chemical composition and localisation of exudate from the vegetative organs of *C. erythrophyllum* was analysed histochemically (Fig. 1). Fresh leaf and stem sections were subjected to a range of stains. Ferric trichloride and N,N-dimethyl-p-phenylene diamine (NADI) reagent were used to evaluated the presence of phenolic compounds and essential oils, respectively. Reagents such as Wagner's and Dittmars were utilised to detect the possible presence of alkaloids.

The emanating micromorphological analysis identified two distinct trichome types, peltate scales and non-glandular trichomes. The head cell count of peltate scales appeared to increase upon leaf maturation and ranged from 8 to 19 cells. In addition, the granulocrine pathway was identified as a possible mode of secretion due to the extensive presence of vesicles, vacuoles and electron dense material within the peltate scales (Fig. 2). Preliminary histochemical analyses indicated the predominant presence of alkaloids, phenolic compounds and essential oils (Fig. 1) within the exudate. These secondary metabolites may serve a crucial ecological role in preventing microbial and herbivory attack. Medicinally, these beneficial compounds may be developed for use in medicinal applications. The overall findings suggest that *C. erythrophyllum* is indeed a plant worthy to be considered for its medicinal properties.

The authors would like to acknowledge the National Research foundation for funding this research.

References:
1. Moura, I. *et al.* (2018) J. Med. Plants Res.12 (2), 15.
2. Lima, G.R. *et al.* (2012) Molecules 17(1), 9142.
3. Mtunzi, F.M. *et al.* (2017) Asian Pac. J. Trop. Med. 10(7), 670.
4. Mawoza, T. and Ndove, T. (2015) Global J. Bio. Agri. Health Sci. 4(1), 105.
5. Martini, N., Katerere, D.R.P. and Eloff, J.N. (2004) J. Ethnopharmacol. 93(1), 207.
6. Fahn, A. (1988) New Phytol. 108(1), 229.
7. Naidoo,Y. *et al.* (2012) Plant Syst. Evol. 298(1), 25.

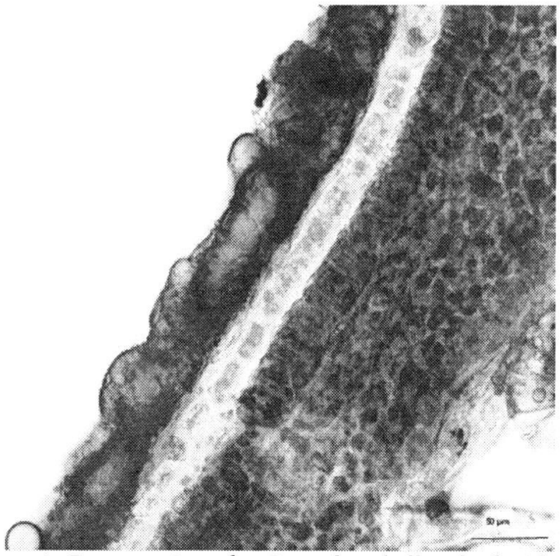

Figure 1. Cross section of a *C. erythrophyllum* leaf with peltate scales, indicating the presence of essential oils, stained using NADI reagent.

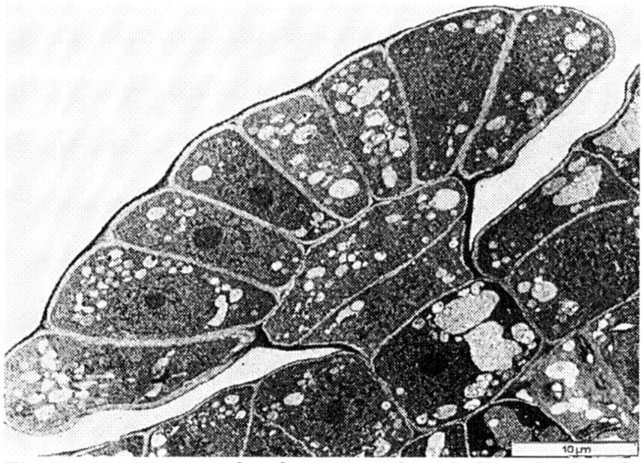

Figure 2. Micrograph of a *C. erythrophyllum* leaf with peltate scales, indicating the presence of vesicles within the stalk and head cells of the pelate scale.

Corresponding author: rahualbantho@gmail.com

THREE NEW SPECIES FROM THE GENUS *Leiotrocha* (PERITRICHIA: CILIOPHORA) FROM SOUTH COAST INTERTIDAL CHITONS

G.P. de Jager and L. Basson

Department of Zoology and Entomology, University of the Free State, Bloemfontein

All members of the genus *Leiotrocha* are ectozoic mobiline symbionts of marine invertebrate hosts, that have been described from various molluscs and polychaetes worldwide. Taxonomic and ecological data of these ciliates in Africa is unfortunately limited, with only two descriptions from limpets on the South African East Coast and Gorée Island, West Africa[1].

This study focused on alpha taxonomy, using scanning electron and compound microscopy, as a means of distinguishing between different species of chiton mobiline symbionts from the intertidal pools of De Hoop Nature Reserve (South Coast).

Five of each randomly selected chiton hosts (*Chiton politus* Spengler, 1797; *Dinoplax gigas* (Gmelin, 1791), and *Acanthochitona garnotti* (Blainville, 1825) were collected from the rocky shores at De Hoop Nature Reserve. Compound microscopy was used to inspect the gills of the hosts for the presence of symbionts. Mobiline samples were fixed in 10% buffered neutral formalin, 4% glutaraldehyde and 99% ethanol, Bouin's solution and on dry smears. Photomicrographs were taken using a Zeiss Axiophot compound microscope and AxioCam ICc 5 for the recommended morphometric analysis[2] of the denticle structure ($AgNO_3$ impregnation method[3]) and nuclear material (Mayer's hæmatoxylin staining[4]). Formalin and glutaraldehyde fixed material was dehydrated, critical point dried, mounted on stubs, sputter-coated with gold and examined at 5kV using a JEOL WINSEM JSM 6400 scanning electron microscope.

Morphometric analysis by means of compound microscopy (Fig. 1) revealed that minimal differences are observed between individual leiotrochans within a population per host, but significant morphological differences between the mobilines found on the different chiton hosts. Electron microscopy further more illustrated differences in pellicle texture (Figs 2A & B), the presence of aboral disc cilia only on one side of the aboral adhesive disc and a unique shape of cilia.

Each of the investigated polyplacophoran genera seems to harbour its own species of mobiline, revealed by the morphometric analysis and verified by ultra structure analysis. All three *Leiotrocha* species possess adhesive disc cilia (Figs 2C & D), a trait specific to this genus. However, these adhesive disc cilia appear to be vestigial, probably residual from the scopular cilia that the sister group (sessile ciliates) use to attach to their substrates. The investigated leiotrochan populations all represent new species that clearly illustrates the high degree of host specificity of chiton symbionts, a tendency that has been deemed irrelevant by others in this field.

In the current scientific milieu which is swiftly being dominated by molecular techniques, this study demonstrates that alpha taxonomy, aided by compound and electron microscopy has stood the test of time and is still the most revealing taxonomic and evolutionary tool.

References:
1. van As, L.L, Basson, L. and van As, J.G. (2017). Acta Protozool. 56, 245.
2. Lom, J. (1958) J. Protozool. 5, 251.
3. Klein, B.M. (1958). J. Protozool. 5, 99.
4. Humason, G.L. (1979) Animal Tissue Techniques. San Francisco, W.H. Freeman and Company.

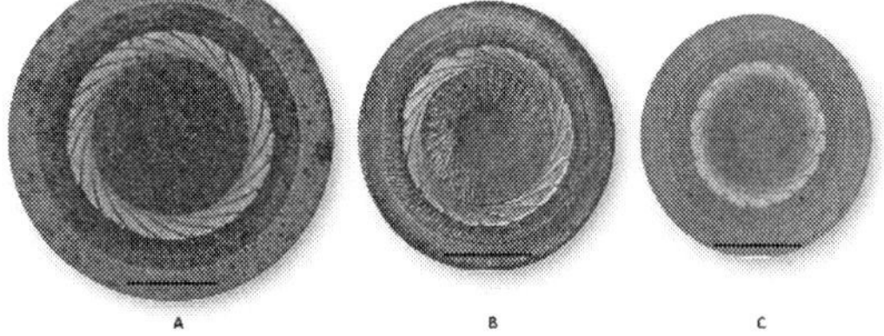

Figure 1. Adhesive discs and comparative sizes of species from A - *Dinoplax gigas* (*Leiotrocha* sp. A), B – *Chiton politus* (*Leiotrocha* sp. B) and C – *Acanthochitona garnotti* (*Leiotrocha* sp. C) (scale: 10µm).

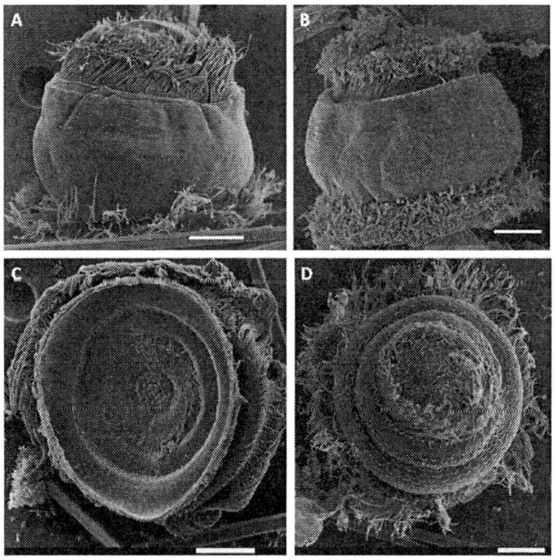

Figure 2. Lateral view of A – *Leiotrocha* sp. C, B – *Leiotrocha* sp. B, highlighting the difference in pellicle striations; aboral view of C – *Leiotrocha* sp. A illustrating the vestigial adhesive disc cilia and D - adoral oral ciliary spiral from *Leiotrocha* sp. B (scale: 10µm).

Corresponding author: bassonL@ufs.ac.za

A MICROSTRUCTURAL ANALYSIS OF A Ti-Ta-Cu-Nb-Zr ALLOY

N.G. Hashe[1], L. Fowler[2], W.E. Goosen[3], S. Norgren[2], N.D. Masia[4], L.A. Cornish[4], C. Öhman-Mägi[2], L.H. Chown[4], S. Magadla[1], K.T. Nape[4], M. Mavundla[4] and L. Spotose[4]

[1]Department of Physics, Nelson Mandela University, Port Elizabeth, [2]Department of Engineering Science, Uppsala University, Uppsala, Sweden, [3]CHRTEM, Nelson Mandela University, Port Elizabeth, [4]School of Chemical and Metallurgical Engineering, and DST-NRF Centre of Excellence in Strong Materials, hosted by the University of the Witwatersrand, Johannesburg

Titanium and its alloys are used in the medical industry as implants because of their biocompatibilities and excellent mechanical and physical properties[1]. The Ti-6Al-4V alloy is used as an implant material due to its high corrosion resistance and excellent biocompatibility[1]. However, there are concerns about the toxicity of aluminium and vanadium[2]. The other concern is the formation of the adherent biofilms around the surface area of the implants[3]. Therefore, the current research is aimed at replacing V and Al with Ta, Nb and Zr to produce a biocompatible alloy with antibacterial[4] and osseointegrative properties that has a similar microstructure to that of an alpha-beta Ti-6Al-4V alloy.

The Ti-10.1Ta-1.7Nb-1.6Zr (wt%) alloy produced by Stenlund *et al.*[5] was studied further. Titanium and Zr sponges were mixed with Ta and Nb wires and compressed into a 7 kg lump, which was melted four times in vacuum in an Electron Beam furnace for homogeneity. The Ti-10.1Ta-1.7Nb-1.6Zr alloy was mixed with 3 mass% Cu and the mixture was melted five times in a custom built arc-furnace. Annealing was done in a sealed quartz-ampoule at 1.333 mbar (after flushing) in a conventional furnace. Heat treatment was done at 980 °C for 10 hours, followed by 798 °C for 6 hours, with a final rapid salt-brine water quench. The annealing temperatures were chosen to obtain the alpha-Ti (hcp) and beta Ti (bcc)[5]. After heat treatment, the samples were prepared metallographically and analysed by scanning electron microscopy (SEM) and X-ray diffraction (XRD).

Backscattered SEM imaging (Fig. 1) shows that the microstructure of the Ti-Ta-Cu-Nb-Zr alloy was lamellar, which are reported to have a high fatigue crack propagation resistance and high fracture toughness[6]. There were small, irregular regions of a much darker phases (indicated by a white arrow) , within the dark phase (indicated by a white star). Energy dispersive spectroscopy (EDX) showed the light contrast phase (indicated by a black arrow) had more Ta and Cu, with less Nb than the dark phase. However, the Nb content was less than 1 mass%, so the results were too near the detection limit to be reliable. The dark phase was identified as alpha Ti since it had more Ti. The small regions of the darkest phase were too small to analyse accurately with EDX.

The XRD patterns only showed alpha-Ti (Fig, 2). The pattern for the heat treated samples also showed apparent preferred orientation, which was surprising considering that the pattern from the as-cast sample was more random. Further work is required to identify the darkest phase, and XRD needs to be repeated with a more suitable X-ray source in order to quantify the phases present in the material.

References:
1. Geetha, M. *et al.* (2009) Prog. Mat. Sci. 54, 397.
2. Niinomi M. and Nakai M. (2011) Int. J. Biomater. 2011, 1.
3. Lorenzetti, M. *et al.* (2015) Appl. Mater. Interfaces 7, 1644.
4. Magadla, S. *et al.* (2019) Euromat 2019 Abstracts, 2288.
5. Stenlund, P. *et al.* (2015) Acta Biomat. 20, 165.
6. Filip, R. *et al.* (2003) J. Mat. Proc. Techn. 133, 84.

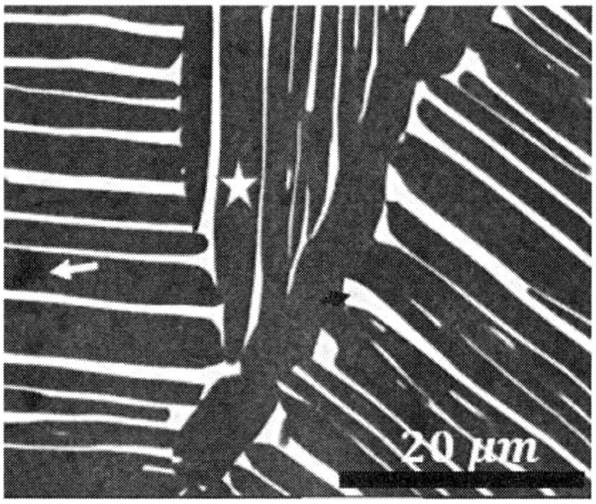

Figure 1. SEM-BSE image showing the typical microstructure of Ti-Ta-Nb-Zr-Cu alloy with three phases.

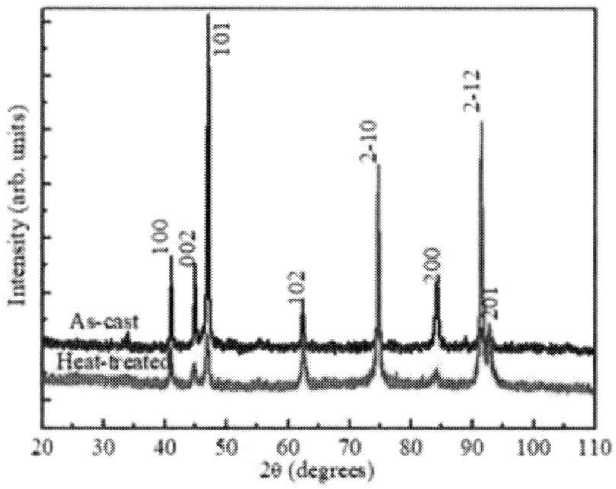

Figure 2. XRD patterns of the as-cast and heat treated Ti-Ta-Cu-Nb-Zr alloy, showing only alpha-Ti.

Corresponding author: nobom.hashe@mandela.ac.za

AN APPLICATION OF CONVOLUTIONAL NEURAL NETWORKS TO STRUCTURE-PROPERTY PREDICTIONS IN LOW-CARBON STEELS

L. Westraadt[1], L. von Wielligh[2] and J.E. Westraadt[3]

[1]Physics, NMU, Port Elizabeth, [2]eNtsa, NMU, Port Elizabeth, [3]Centre for HRTEM, NMU, Port Elizabeth

One of the major challenges in characterising materials for industrial applications is the need to *a priori* define the microstructural features that must be measured[1]. In reality, material systems are often highly heterogeneous, and many material properties are influenced by unknown details of this heterogeneity in the material structure[1]. Recent advances in machine learning are providing novel solutions in this regard. In particular, convolutional neural networks (CNNs), a deep learning approach in computer vision, can learn from micrographs without feature engineering[1]. The aim of this study was to test the feasibility of using CNNs to predict structure-property relationships in low-carbon steels.

WeldCore®, a novel sampling and repair technique for in-situ sampling of high pressure steel components was used to remove site specific core samples for the investigation. Small punch creep (SPC) testing was performed on 8 mm disk specimens, 0.5 mm thick, extracted from the core samples. Etched (2% Nital) and polished cross-sections of each sample were then investigated using visible light microscopy to determine the microstructure of the sample. A total of 90 data tuples were generated to develop the CNN. Each tuple consisted of a 128×128 micrograph with its corresponding time to rupture.

Contemporary deep learning approaches require a large amount of data and computational resources[1]. In contrast, the amount of microstructural data available for materials studies is small[1]. Recent applications of transfer learning to the study of materials systems[2-5] have successfully circumvented this problem. In this study, a transfer learning approach similar to the one employed by Li and colleagues[4] was employed. The architecture of the neural network was constructed using the output of the pooling_2 layer in the VGG-19 model[6]. The latter was flattened, followed by two fully connected layers of size 2048 pixels and 1024 pixels with Rectified Linear Unit (ReLU) activations. The weights of the transferred layers were initialised using the pre-trained weights in the VGG-19 model while the remaining weights were initialised using the default method in Keras. Data was split 80%-20% for training and testing, respectively. The Adam optimiser (learning rate = 0.0005, beta1 = 0.9, beta2 = 0.99) was used to fine-tune the parameters. The mini-batch size was set to 16 and the number of epochs was 2,000. Regularisation was not employed as it had limited impact on model performance.

Fig. 1(a) and (b) show representative input microstructures corresponding to low and high rupture times, respectively. For medium to high rupture times, the microstructures consisted of alternative bands of ferrite (red) and pearlite (blue). Fig. 2 shows the predicted rupture time as a function of the target values for both the training (blue) and test (red) sets. Model predictions were also compared to a simpler model based on one-point statistics[1] (green). The latter was based on a linear regression analysis of the relationship between rupture time and the volume fraction ratio of pearlite to ferrite. The performance of the CNN model was comparable to the simpler model, with both models tending to under predict rupture times for well performing samples. This bias in the models could be attributed to missing information such as differences in the mechanical properties of the pearlite phase, input chemistry (such as solid solution strengthening) and grain orientation texturing effects.

References:
1. Simmons, J.P. *et al.* (2019) Statistical Methods for Materials Science: The Data Science of Microstructure Characterization. Boca Raton, CRC Press.
2. Cecen, A. *et al.* (2018) Acta Materialia 146, 76.
3. DeCost, B.L., Francis, T. and Holm, E.A. (2017) Acta Materialia 133, 30.
4. Li, X. *et al.* (2018) Sci. Rep. 8(1), 13461.
5. Yang, Z. *et al.* (2018) Comp. Mater. Sci. 151, 278.
6. Simonyan, K. and Zisserman, A. (2014) arXiv:1409.1556.

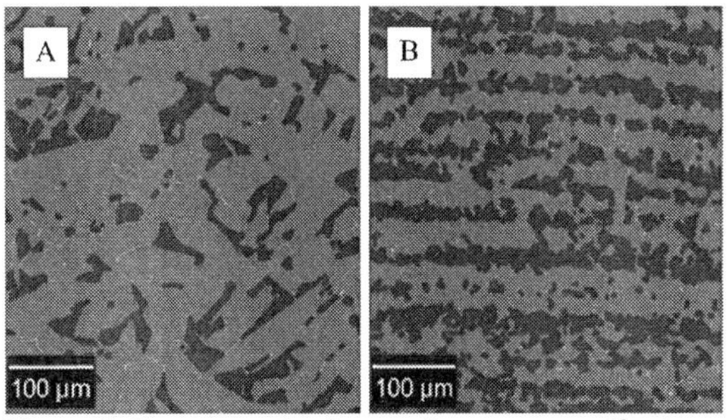

Figure 1. Representative input microstructures (ferrite – red, pearlite - blue) corresponding to (a) low and (b) high rupture times.

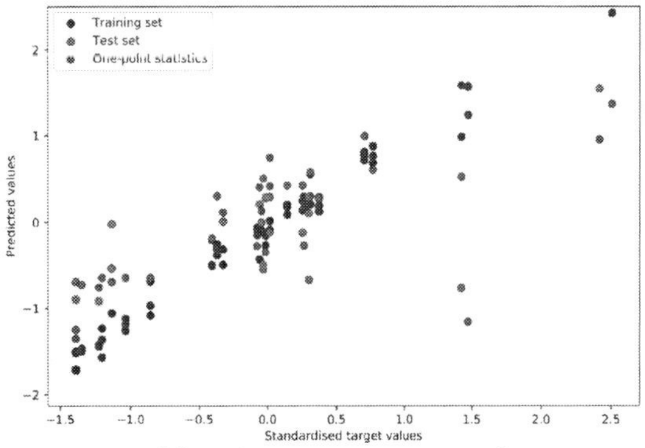

Figure 2. Model predictions versus target values.

Corresponding author: lindsayw@mandela.ac.za

CHARACTERISATION OF CHAIN-LIKE CARBON NANO-ONIONS USING ELECTRON MICROSCOPY AND OTHER RELATED TECHNIQUES

T.H. Mongwe[1], N.J. Coville[1], M.S. Maubane-Nkadimeng [1,2] and M.A. Mamo[3]

[1]DST-NRF Centre of Excellence in Strong Materials and the Molecular Sciences Institute, School of Chemistry, University of the Witwatersrand, Johannesburg, [2]Microscopy and Microanalysis Unit, University of the Witwatersrand, Johannesburg, [3]Department of Applied Chemistry, University of Johannesburg, Johannesburg

The fullerene-like carbon nanomaterials (sizes: <100 nm) herein reported as chain-like carbon nano-onions (CLCNOs), are of significant interest and have captured researchers interest in the scientific field of carbon nanomaterials[1]. This material is made of quasi-spherical nanoparticles (QSNPs) possessing unique chemical and physical characteristics. Furthermore, this interesting form of carbonaceous nanomaterial is termed carbon nano-onions due to the cross sectional appearance of QSNPs that exhibit concentric multi-layers resembling the structure of an onion[2]. The characteristics exhibited by this material make them candidates for multiple possible applications. The material can be applied in various biological and environmental applications as well as in different electronic devices; e.g. biological imaging, various sensing devices, capacitors, and lithium batteries[3,4].

In this study, dense (solid) CLCNOs were synthesized using a controlled flame assisted technique known as the oil-wick method. During the synthesis, the controlled flame was produced using olive oil (Hercules product) as the fuel. The black carbonaceous powder from the flame was trapped and collected using a flat brass plate (140×140×5 mm). This method follows a combustion process and it entails the breaking down of the liquid fuel molecules into different radicals, including carbon radicals, which further nucleate and grow to form aggregated quasi-spheres of carbon nanomaterial with concentric multi-layered nanoparticles. In principle, the size of conglomerated carbon nano-onions can be regulated by the nucleation and accretion processes of the carbon radical species formed in the flame.

The structural morphology and characteristics of CLCNOs were obtained using scanning electron microscopy (SEM) and transmission electron microscopy (TEM). Further characterization of CLCNOs was conducted by using other analytical techniques; Brunauer-Emmett-Teller (BET) surface analyses, Fourier transform infrared (FTIR) and Raman spectroscopy, thermo-gravimetric analysis (TGA), and powder X-ray diffraction (PXRD).

The fundamental analysis for this material included electron microscopy analysis. SEM and TEM micrographs of as-synthesized CLCNOs particles are shown in Fig. 1. The as-synthesized CLCNOs particles have quasi-spherical morphology (average particles size: ~58 ± 3.4 nm) as shown in Fig. 1(a). This sample consists of carbonaceous nanoparticles that exhibit different regions of carbon concentric layers. The concentric layers have an estimated inter-layer separation of ~0.35 ± 4.6 nm as shown in Fig. 1(b). The BET analysis revealed that the surface area of CLCNOs is ~63.6 m^2/g with pore volume and pore sizes of about 0.1 cm^3/g and 5 nm respectively. The only carbon phase ((002 and 004) plane) peaks were also observed from the PXRD pattern which confirmed that the nanomaterial are composed of carbon based nanoparticles.

Further characterisation has shown that the as-synthesized material have graphitic nature. This was revealed by Raman spectrum which showed that the G-band intensity (indicting the in-plane vibration mode of sp^2 carbons) of CLCNOs, is higher than the D-band intensity (indicating the disruptions in sp^2 bonding, such as dangling bonds, sp^3 bonding, vacancies, and carbon rings other than hexagon)[5,6]. Moreover, the material is thermally stable up to ~500 °C as confirmed by TGA analysis. The FTIR results showed the existence of different CLCNOs surface functional groups.

The synthesis of this allotrope of carbon nanomaterials yields cleaner quasi-spherical material with particles sizes that are less than 100 nm. Hence, these nanomaterial exhibit special chemical and physical properties which make them candidates for various applications.

References:
1. McDonough, J.K. and Gogotsi, Y. (2013) Electrochem. Soc. Interface. 1(1), 62.
2. Kausar, A. (2017) Fuller. Nanotub. Car. N. 25(10), 110.
3. Zeiger, M. et al. (2016) J. Mater. Chem. A. 4(1), 3173.
4. Mykhailiv, O., Zubyk, H., Plonska-Brzezinska, M.E. (2017) Inorg. Chim. Acta. 468(1), 50.
5. Tripathi, K.M. et al. (2016) RSC Adv. 6(1), 37320.
6. Mongwe, T.H. et al. (2018) J .Diamond. 90(1), 139.

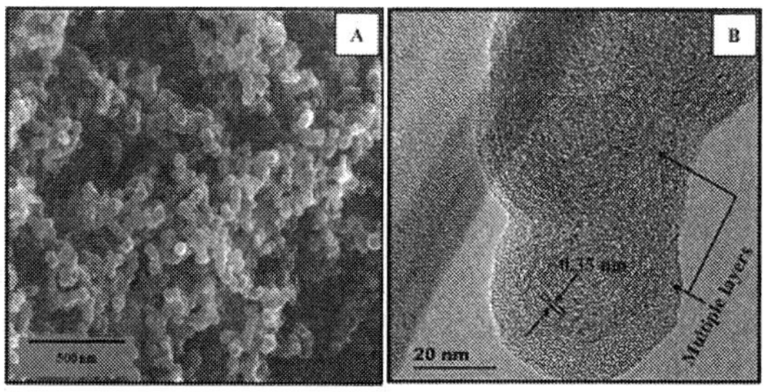

Figure 1. (a) SEM micrograph of Quasi-spherical as-synthesized CLCNOs and (b) TEM micrograph of multi-layered as-synthesized CLCNOs.

Corresponding author: 1585603@students.wits.ac.za

Co CATION VALENCE STATE AND STRUCTURAL STABILITY OF $NiFe_2O_4@Co_3O_4$ NANOPARTICLES DURING IN-SITU TEM REDUCTION

V. Shumane[1], E.J. Olivier[1], A. Govender[2] and S.J. Haigh[3]

[1]Centre for HRTEM, Physics Department, Nelson Mandela University, Port Elizabeth, [2]Group Technology, Sasol South Africa (Pty) Ltd, Sasolburg, [3]School of Materials, University of Manchester, Manchester, UK

Cobalt-based catalysts are widely used for the low-temperature Fischer-Tropsch reaction (LTFT) to produce middle distillates and long-chain linear hydrocarbons. Despite their high cost, these catalysts are preferred because of their high activity, high selectivity to linear paraffin's, low water-gas shift activity and stability to deactivation by water. In order to justify the higher costs associated with cobalt catalysts, various efforts are made in order to ensure that it operates for a longer period and have a good activity[1]. Regeneration of deactivated cobalt catalysts has been examined as a method to improve the economics of the overall cobalt-based Fischer-Tropsch process[2]. Another method to lower the cost of the cobalt-based catalyst is to replace the non-catalytically active cobalt in the catalyst with a cheaper alternative which would result in the formation of a core-shell structure[3]. The applicability in the use of such core-shell structures within the proposed industrial processes, however, depend on the structural stability of the catalyst after activation and during reaction.

In this present work core-shell catalyst precursors with nickel ferrite, $NiFe_2O_4$ as core and cobalt oxide as shell was used as core-shell system to study the structural evolution and Co valence state at various steps (temperatures and time) during an in-situ gas TEM reduction investigation. The core-shell catalyst precursors were prepared using a simple precipitation route using cobalt carbonate, ammonium carbonate and ammonia solution as reactants to produce a Co_3O_4 shell around a $NiFe_2O_4$ core[3]. The in-situ TEM analysis was carried out on a FEI Titan G2 80 – 200 (S)TEM ChemiStem™ operating at 200 kV. The instrument was equipped with a Gatan GIF Quantum ERSTM spectrometer utilised for EELS spectrum image mapping. The gaseous e-Cell system used in the FEI Titan instrument was a Protochips Atmosphere holder with a customized low penumbra geometry beryllium lid. During the measurements, the e-Cell was filled with pure hydrogen at nominally one bar pressure and heated at temperatures ranging between 180 °C to 280°C for hold times of 1hr at each temperature set point.

The acquired EELS spectrum image data obtained after each reduction step was processed to extract the distribution of Fe, Co and Ni as well as Co valence state of the core-shell nanoparticles. The Co cation valence state after each reduction step was determined by extracting the ratio of the Co-L3 to Co-L2 peak intensity present in the Co-L near-edge fine structure[4]. Figure 1 is an example of an as-prepared $NiFe_2O_4@Co_3O_4$ showing the elemental distribution of Fe, Co and Ni as well as a Co-L3 to Co-L2 peak height ratio mapping across the region. Figure 1 C shows a representative EELS spectrum of the Fe-L, Co-L and Ni-L used for the analysis.

From the findings it was determined that the complete reduction of Co to zero-valent state was achieved for the $NiFe_2O_4@Co_3O_4$ core-shell system after 2hrs reduction at 180°C with the core-shell structure remaining intact. With temperatures of 230°C and above reduction of Co to the zero-valent state was accompanied with a surface restructuring of the nanoparticles with diffusion of Fe into Co regions observed.

References:
1. Khodakov, A.Y., Chu, W. and Fongarland, P. (2006) Chem. Rev. 107, 1692.
2. Saib, A.M. *et al.* (2010) Catal. Today 154, 271.
3. Govender, A. *et al.* (2016) Mater. Charac. 121, 93.
4. Wang, Z.L., Yin, J.S. and Jiang, Y.D. (2000) Micron 31(5), 571.

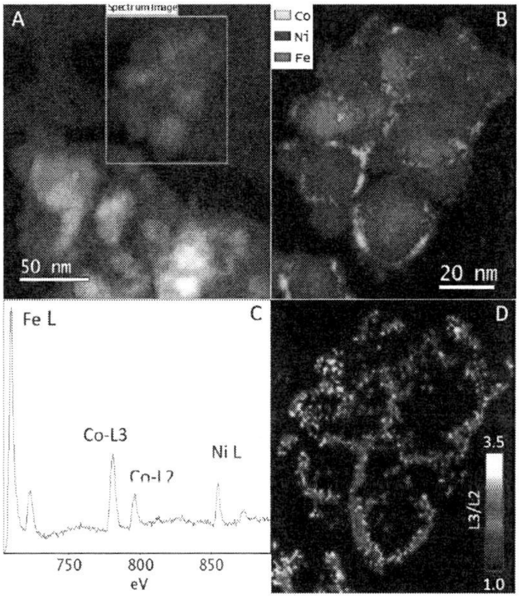

Figure 1. As-prepared $NiFe_2O_4@Co_3O_4$ showing the elemental distribution (B) of Fe, Co and Ni as well as a Co L3/L2 ratio mapping (D) across the region. Figure 1 C shows a representative EELS spectrum of the Fe-L, Co-L and Ni-L used during the analysis. Figure 1 A is ADF STEM image of the area studied.

Corresponding author: shumanev2@gmail.com

CORROSION OF nc-TiZrN COATED ZIRLO IN HIGH-TEMPERATURE OXYGEN

H.S. Nyembe[1], J.H. Neethling[1], A. Janse van Vuuren[1] and V.V. Uglov[2]

[1]Centre for HRTEM, Nelson Mandela University, Port Elizabeth, [2]Department of Solid State Physics, Belarusian State University, Belarus

Zirconium alloys are used as nuclear fuel cladding in water cooled nuclear power plants. In the reactor, the Zr alloy undergoes waterside corrosion and absorbs hydrogen. Corrosion and hydrogen absorption in Zr alloys degrade the properties of the cladding and limit the lifetime and reloading cycles of fuel in nuclear power stations[1]. To reduce corrosion, a proposed solution is the application of a corrosion-resistant coating on the outer surface of Zr alloy tubes. Nanocrystalline TiZrN coatings have promising corrosion resistance characteristics[2] and have been proposed as candidate coatings to improve the corrosion resistance of Zr claddings. In this study, the microstructure, composition and corrosion resistance of nc-TiZrN coatings on ZIRLO is investigated. High-temperature corrosion tests in an oxygen environment were performed to determine oxidation kinetics.

The nc-TiZrN coatings were deposited onto flattened ZIRLO tube-sections with a cathodic arc vapour deposition technique. Coated and uncoated ZIRLO samples were sectioned into smaller sizes for corrosion testing. Corrosion experiments were conducted in a horizontal quartz tube furnace under flowing oxygen. The samples were heat treated at 500 °C for 6 to 20 hours. The unexposed and exposed samples were subsequently sectioned, mounted and polished for characterization. The characterization was performed using JEOL JSM 7001F SEM and JEOL 2100 TEM both equipped with Oxford EDS detectors. A Helios NanoLab FIB-SEM was used to extract TEM lamellae. MIPAR software was used to measure the average thickness of the coating and oxide layers.

The as-produced TiZrN layers are 3.6 µm thick with a small number of observable Ti-rich and Nb-rich macroparticles. TiZrN exhibits a single-phase solid solution of $Ti_{0.42}Zr_{0.58}N$ with a NaCl-type crystal structure and columnar nanocrystal grains which have <111> and <200> preferred orientations as shown in Fig. 1. The thickness of the TiZrN and ZIRLO consumed by oxidation as a function of time in log-log scale and the oxidation rate exponent (n) are shown in Fig. 2. ZIRLO exhibited larger consumed thicknesses than TiZrN. The corrosion rate of TiZrN is much lower compared to ZIRLO. The linear kinetics (n = 0.9) obeyed by TiZrN indicate the existence of a non-protective TiZr-oxide. Parabolic kinetics (n = 0.5) coupled with a Zr-oxide with dense microstructure was observed for ZIRLO. The microstructure of the TiZr-oxide of the coated sample exposed for 20 h is shown in Fig. 3. The TiZr-oxide layer is shown to be porous which allows rapid diffusion of oxygen ions through the oxide during corrosion. This is consistent with the linear kinetics for TiZrN. Although the oxidation resistance of the nc-TiZrN coating is higher than that of ZIRLO, linear oxidation kinetics combined with the microstructure of the TrZr-oxide is not well suited to long-term service in the harsh nuclear reactor environment.

References:
1. Bae, H. *et al.* (2015) Corros. Sci. Technol. 14(6), 301.
2. Tang, C. *et al.* (2017) Corros. Rev. 35(3), 141.

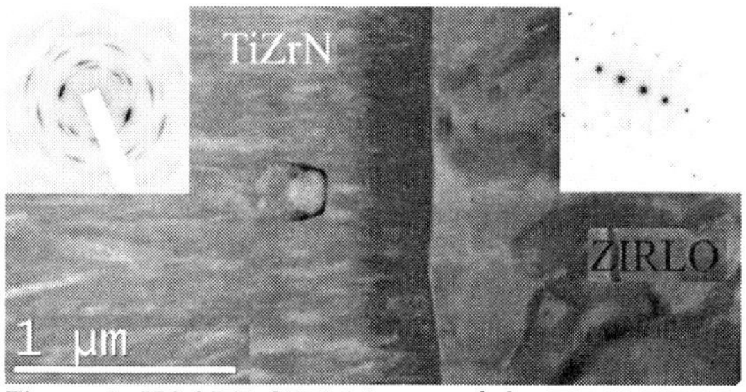

Figure 1. HAADF STEM image of the TiZrN-ZIRLO interface and the corresponding SAD patterns as inset.

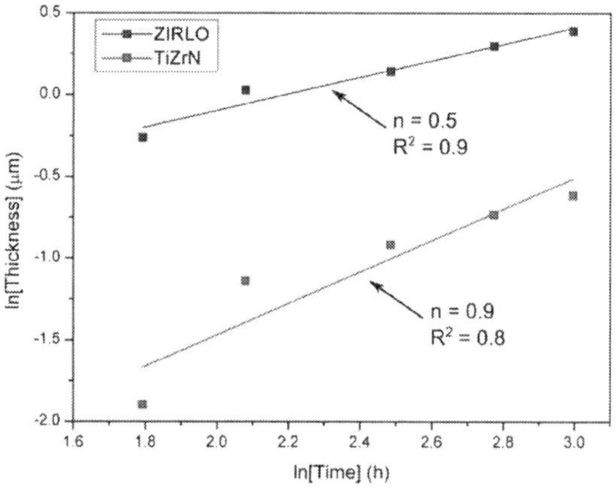

Figure 2. Metal consumed by corrosion as a function of time in log-log scale.

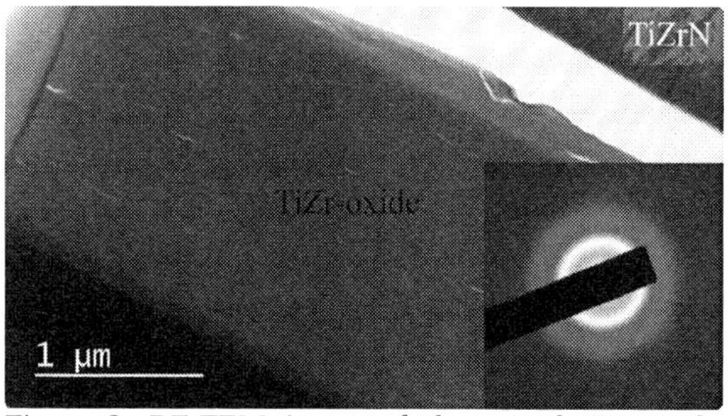

Figure 3. BF-TEM image of the mixed TiZr-oxide formed after corrosion for 20 hours and the corresponding SAD pattern as inset.

Corresponding author: s216800609@mandela.ac.za

DETERMINATION OF STRUCTURAL PARAMETERS OF V/MgO CATALYSTS PREPARED USING SOLUTION COMBUSTION SYNTHESIS

E.J. Olivier[1], P. Ntola[2], H.B. Friedrich[2], A.S. Mahomed[2], S. Singh[2] and A. Govender[3]

[1]Centre for HRTEM, Physics Department, Nelson Mandela University, Port Elizabeth, [2]School of Chemistry, University of KwaZulu Natal, Durban, [3]Group Technology, Sasol South Africa (Pty) Ltd., Sasolburg

Vanadium magnesium oxide (V/MgO) is highly desirable in its use as a catalyst for the oxidative dehydrogenation (ODH) of alkanes[1]. It yields high activity with a reported ability to form minimum oxygenates and/or cracking products during reaction, with adequate selectivity towards desired ODH products[1]. Preparation of V/MgO catalysts has typically been done using a variety of synthesis methods which include impregnation using ammonium metavanadate solutions, and co-precipitation methods[2,3]. In this paper, experimental findings in studying the structural morphology of V/MgO catalysts synthesized using solution combustion synthesis (SCS) is reported. The use of SCS for synthesis allows for a novel one-step synthesis of V/MgO catalysts, providing benefits in synthesis time, reproducibility and cost.

Different V/MgO catalysts were prepared by combustion (400 ^{0}C in a muffle furnace) of stoichiometric solutions (1:1 fuel to oxidant) of magnesium nitrate hexahydrate ($MgNO_3.6H_2O$) and ammonium metavanadate (NH_4VO_3), as metal precursors, with different fuels (urea, citric acid, hydrazine hydrate, oxalyl dihydrazide and glycine). The as-prepared materials were characterized using aberration-corrected TEM imaging and diffraction and electron energy loss spectroscopy (EELS) spectrum imaging (Fig. 1) with related techniques. The acquired data for each catalyst prepared was processed to extract quantitative information describing the abundance, composition, distribution and structural morphology of V/MgO phases formed. Obtained parameters were subsequently used as inputs for the refinement of thermodynamic models describing the formation of SCS derived V/MgO catalyst for each fuel-oxidant system.

Figure 1 shows elemental distribution (Mg, O, V) and compositional (V) maps of an area in a V/MgO catalyst (prepared using Urea as fuel) generated from an EELS spectrum image dataset acquired from the region. Figure 2 (left) shows a dark field TEM image of an area in the V/MgO catalyst (Urea) obtained using hollow-cone diffractive imaging centered on vanadate interplanar spacings in the diffraction plane. Figure 2 (right) is a selected area diffraction (SAD) pattern (inverted) used to identify the presence (diffraction spots inside red circle) of magnesium vanadates (active phase) in the as-prepared material.

From the quantitative extraction of structural parameters from the data obtained for each catalyst synthesized the following findings are reported. The predominant phases formed were MgO (periclase) and Mg_3V2O_8 (orthovanadate) consistent with previous X-ray diffraction data. Crystallites of vanadates formed were smaller and less abundant than MgO crystals in all cases studied. Average crystallite size of each phases is qualitatively correlated to the adiabatic flame temperature for each fuel used. Vanadate formation was not observed for all fuels studied. Areas of V enrichment were observed for all catalysts studied with the incorporation of 1 to 15 at % V into MgO phases observed. The degree of incorporation of V into MgO was different for different fuels used and also varied between MgO crystals within each catalyst.

References:
1. Elkhalifa, E.A. and Friedrich, H.B. (2011) Catalysis Letters 141, 554.
2. Soenen, V., Herrmann, J.M. and Volta, J.C. (1996) Journal of Catalysis 159, 410.
3. Dejoz, J.M. *et al.* (1999) Applied Catalysis A: General 180, 83.

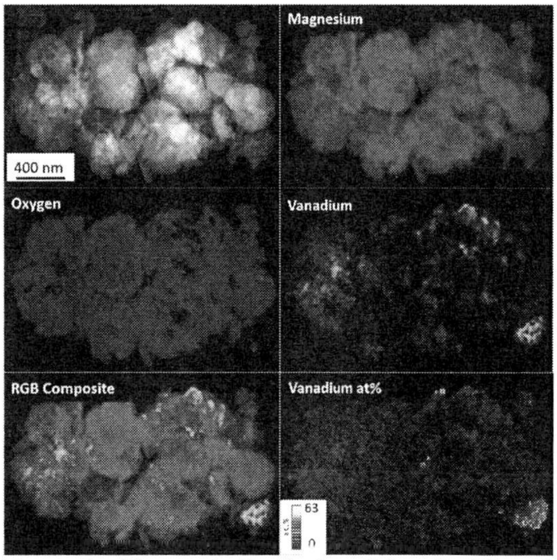

Figure 1. Elemental distribution (Mg, O, V) and compositional (V) maps of a Urea V/MgO catalyst

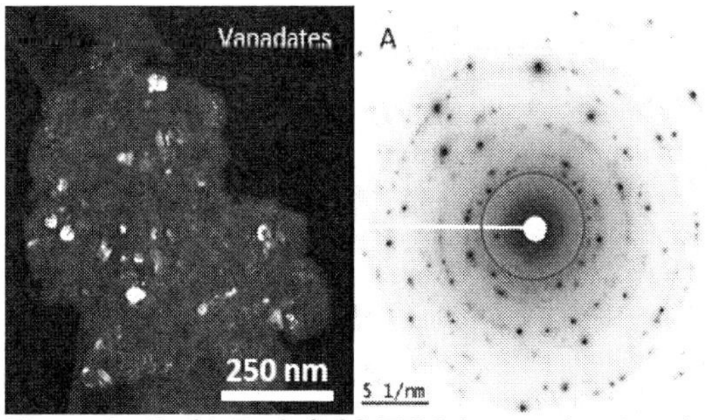

Figure 2. (Left) Dark-field TEM image showing the presence and distribution of vanadates in a Urea V/MgO catalyst. (Right) SAD diffraction pattern of a Urea V/MgO catalyst.

Corresponding author: jolivier@mandela.ac.za

DOPING OF ZrO2 LAYER ON ZIRLO FOR PREVENTION OF HYDROGEN PICKUP

S. Ngongo[1], A. Janse van Vuuren[1], J.H. Neethling[1] and V.V. Uglov[2]

[1]Centre for HRTEM, Department of Physics, Nelson Mandela Metropolitan University, Port Elizabeth, [2]Belarusian State University, Minsk, Belarus

Zirconium alloys such as ZIRLO are widely used as fuel cladding in water cooled nuclear reactors. When a loss of coolant accident occurs, the temperature of the fuel tubes will increase resulting in enhanced oxidation of the zirconium alloy fuel tube surface and hydrogen pick-up by the fuel tube. The hydrogen atoms diffuse into the fuel tube and forms brittle zirconium hydrides[1]. Theoretical predictions by Youssef and co-workers[2] indicated that by doping the oxide layer on the zirconium alloy with chromium, the solubility of hydrogen in ZrO_2 is greatly decreased which reduce the pick-up of hydrogen by the zirconium fuel tube.

In this study, compression plasma flow (CPF) was used to produce a chromium doped, oxidized ZIRLO surface layer. CPF is generated by quasi-stationary plasma accelerators with own magnetic field. Such plasma flows are characterized by long life-time (about 100 µs) and high energy density absorbed by the target (from 10 to 100 J/cm^2). When CPF interacts with the surface, the top layer is melted and a mixing process takes place in the layer. If a metal coating is deposited on the surface of the treated sample, the CPF impact will provide mixing of both coating and the substrate. A chromium coating with a thickness of 1 µm was deposited on the oxidized ZIRLO surface. The CPF treatment was done in a nitrogen atmosphere (400 Pa pressure). The chromium alloyed ZIRLO samples were sectioned using a diamond wire saw and mounted on a stub. A Helios NanoLab FIB SEM was used to cut transmission electron microscopy (TEM) lamellae from specific areas of interest. The TEM lamellae were investigated in a JEOL 2100 LaB_6 TEM operated at 200 kV.

Figure 1 is a BSE SEM micrograph of a CPF treated ZIRLO surface. The micrograph shows the surface morphology of the chromium alloyed ZIRLO oxide layer. Figure 2 (a) is a cross-sectional TEM micrograph of the layer produced by CPF treatment while the energy dispersive X-ray spectrometry (EDS) elemental maps of O, Cr and Zr after the CPF treatment are shown in (b).

The results of this investigation indicate that the alloying process of the oxide layer by the CPF method was successful. Results of hydrogen desorption from the chromium doped oxidized ZIRLO surface layers were found to be in agreement with the theoretical predictions by Youssef and co-workers[2].

References:
1. Motta, A.T. and Chen, L.Q. (2012) J. Min. Met. Mater. Soc. 64, 12.
2. Youssef, M. *et al.* (2016) Phys. Rev. Applied. 59, 014008.

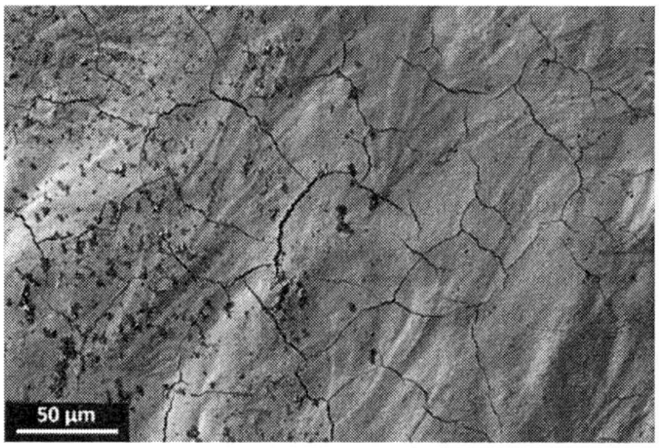

Figure 1. BSE SEM micrograph surface CPF treated ZIRLO.

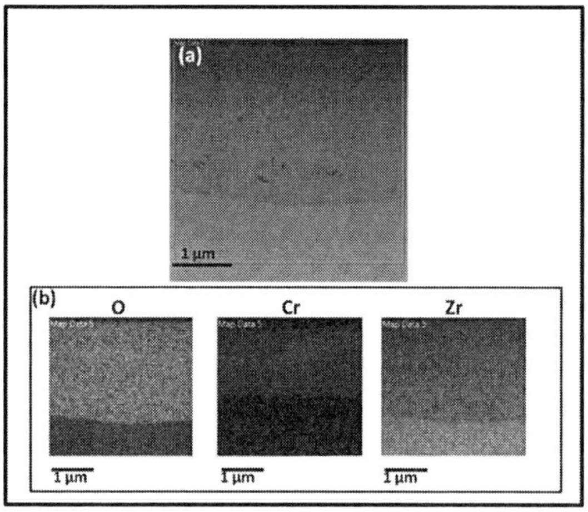

Figure 2. (a) Cross-sectional TEM micrograph of the CPF treated ZIRLO surface. (b) EDS elemental maps of O, Cr and Zr after CPF treatment.

Corresponding author: s214297012@mandela.ac.za

EFFECT OF DIFFERENT COPPER ADDITIONS TO Ti–10.1Ta–1.7Nb–1.6Zr

K. Dyal Ukabhai[1], L. Fowler[2], K.T. Nape[1], M. Mavundla[1], L. Spotose[1], S. Magadla[3], N.D. Masia[1,4], D.E.P. Klenam[1], N.G. Hashe[3], C. Öhman Mägi[2], S. Norgren[5], L.H. Chown[1] and L.A. Cornish[1]

[1]Sch. Chem. & Met. Eng. and DST-NRF CoE Strong Materials, hosted by U.Witwatersrand, [2]Dept of Eng. Sciences, Ångström Lab. Uppsala University, Sweden, [3]Dept Physics, Nelson Mandela University, Port Elizabeth, [4]Advanced Materials Division, Mintek, Randburg, [5]Sandvik Coromant R&D, Stockholm, Sweden

Commercially pure titanium is popular for bone-anchored dental devices, while the stronger Ti-6Al-4V alloy is preferred for higher load applications. However, Al and V are potentially toxic, giving reason to develop new (α+β) alloys without these elements, but with similar proportions of the two phases: (αTi) (brackets denote the solid solution based on Ti with that structure) and (βTi), for similar mechanical properties to Ti-6Al-4V. Stenlund *et al.*[1] undertook computer calculations to derive "TTNZ", an (α+β) alloy with similar phase proportions to Ti-6Al-4V, but without Al or V: Ti-10.1Ta-1.7Nb-1.6Zr (in wt%). The alloy was manufactured from Ti and Zr sponge, and Ta and Nb wire, which were mixed, compressed, then melted in vacuum in an electron beam furnace to generate a homogeneous alloy.

Around 7% of patients with dental implants suffer from peri-implantitis within 10 years due to infection[2]. Copper is known to have antimicrobial properties[3]. Liu *et al.*[4] studied sintered Ti–*x*Cu alloys (*x* = 2, 5, 10 and 25 wt%) using XRD and SEM, and the antibacterial activity was assessed. The Ti_2Cu phase was found in all alloys. A Cu-rich phase formed in alloys with >5wt% Cu, which were also the best alloys for antibacterial activity.

This study used TTNZ samples made in the same way as Stenlund *et al.*[1] with nominal 1, 3, 5 and 10 wt% Cu added. The samples were prepared metallograpically, and studied using XRD and SEM-EDX after Vickers hardness tests ($HV_{0.1}$). XRD showed only (αTi) and (βTi) in the 0 and 1 wt% Cu TTNZ-Cu alloys, but the 3, 5 and 10 wt% Cu alloys had increasing amounts of Ti_3Cu[5], instead of the more commonly-reported Ti_2Cu[4]. SEM EDX gave the copper content of Ti_3Cu as 28.1±0.3 to 29.7±1.1 at.% Cu. Increased Cu changed the microstructures from mainly (βTi) with small amounts of precipitated (αTi) for 0 and 1 wt% Cu (Fig. 1) to more equal amounts of the two phases with 3 wt% Cu (Fig. 2). Further increase in Cu increased the Ti_3Cu proportions. The 10 wt% Cu alloy had a totally different microstructure (Fig. 3), with fewer and shorter (αTi) laths, and Ti_3Cu precipitated on the prior (βTi) grain boundaries. Hardness increased with copper addition up to 5 wt% Cu, but 10 wt% Cu had lower hardness due to its different microstructure. The hardnesses were: 145±8, 182±6, 200±10, 201±8 and 178±8 $HV_{0.1}$ for 0, 1, 3, 5 and 10 wt% Cu respectively. Although most of the hardness indentations were square, Fig. 2 shows there was less plastic deformation of the harder Ti_3Cu phase. The lower hardness of the 10 wt% Cu alloy indicated too much copper had been added.

References:
1. Stenlund, P. *et al.* (2015) Acta Biomat. 20, 165.
2. Derks, J. *et al.* (2015) J. Dent. Res. 94(3), 44S.
3. Grass, G., Rensing, C. and Solioz, M. (2011) Appl. Environ. Microbiol. 77, 1541.
4. Liu, J. *et al.* (2014) Mat. Sci. & Eng. C35, 392.
5. Canale, P. and Servant, C. (2002) Z.Metallkd. 93(4), 273.

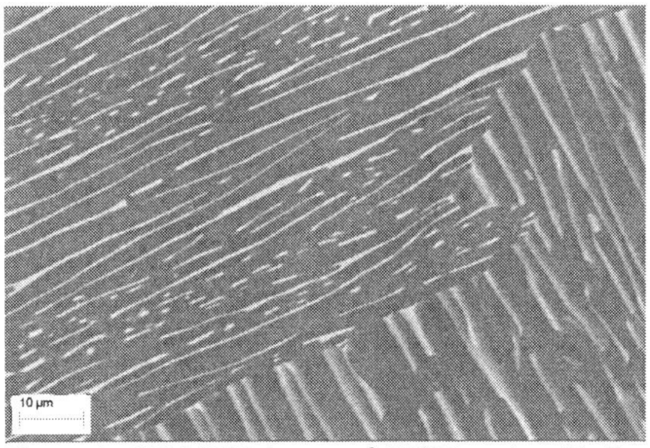

Figure 1. SEM-BSE image of TTNZ + 1 wt% Cu, showing light contrast (βTi) and dark contrast (αTi).

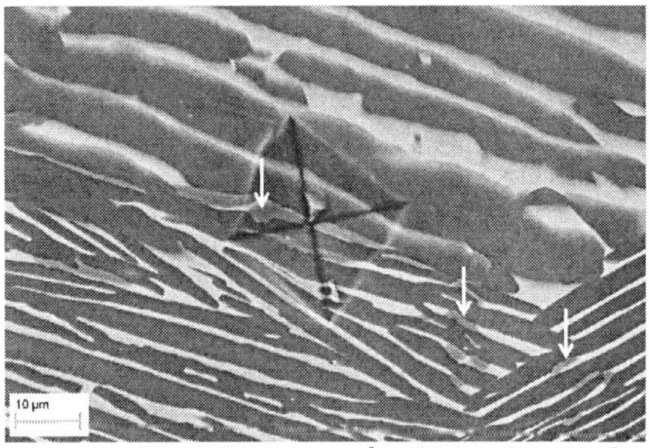

Figure 2. SEM-BSE image of TTNZ + 3 wt% Cu, showing a hardness indentation with light (βTi), dark (αTi) and medium contrast Ti_3Cu (shown by arrows).

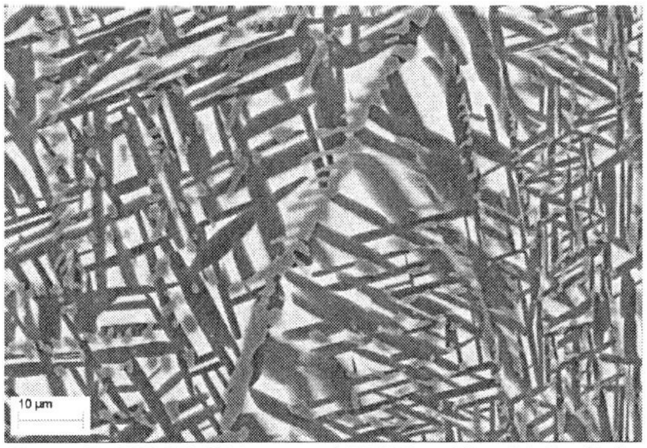

Figure 3. SEM-BSE image of TTNZ + 10 wt% Cu: light (βTi), dark (αTi) and medium contrast Ti_3Cu.

Corresponding author: lesley.cornish@wits.ac.za

EFFECT OF MILLING TIME ON DISPERSION OF SILICON CARBIDE PARTICLES IN ALUMINIUM MATRIX

M. Lesufi , S.O. Akinwamide , P. Leremi , N. Kumalo and P.A. Olubambi

Centre for Nanomechanics and Tribocorrosion, School of Mining, Metallurgy and Chemical Engineering, University of Johannesburg, Johannesburg

Due to the high demand for light materials with enhanced properties, the upgrade of techniques to improve the production of high-performance composite materials is of great interest. Aluminium (Al) metal matrix composites have been developed as an alternative over other conventional engineering materials[1]. Some notable attributes identified with this class of material include improved mechanical properties and excellent distribution of the reinforcement phase in the aluminium matrix[2]. This study investigated the dispersion of silicon carbide (SiC) particles within the aluminium matrix at different milling times.

In this study, 2 vol. % SiC reinforcement was dispersed in the matrix of aluminium, with the addition of stearic acid as a process control agent using high energy ball milling. The milling was performed for 6 and 12 hrs respectively at a constant rotating speed of 150 rpm with a ball to powder ratio of 5:1. The morphologies of the starting (aluminium and silicon carbide) and milled powders were investigated using a backscattered electron detector (BSD), on a field emission scanning electron microscope (FESEM).

Figure 1 present the SEM micrographs of the aluminium and silicon carbide starting powders respectively. The Al particles are observed to be spherical, while the silicon carbide particles show a polyhedric structure. Figures 2 and 3 present the micrographs of powders milled for 6 and 12 hrs respectively. The SiC particles were observed to be embedded within the aluminium matrix with homogeneous distribution as indicated with the red arrows. However, with the Al-SiC powders milled for 12 hours, a flattened sheet-like morphology which resulted from longer milling time was observed. The flat shape was ascribed to the ductile nature of aluminium.

Furthermore, 12 hrs milling time subjected the powders to crushing and work hardening, which resulted from the impact of the grinding balls. The addition of stearic acid was also observed to prevent cold welding between powders and milled Al-SiC powders. This was observed with the pre-tests ran with and without the presence of stearic acid. This study showed that the morphology of the milled powders depended on the milling time, as silicon carbide particles were observed to be homogeneously dispersed in the matrix of aluminium after 12 hrs milling time. Further research is recommended on the study of morphologies obtained from the dispersion of silicon carbide particles in the matrix of aluminium at 18 and 24 hrs milling time.

References:
1. Liu, X. *et al.* (2017) Materials Characterization 133, 122.
2. Woźniak, J. *et al.* (2014) Archives of Metallurgy and Materials 59(4), 1493.

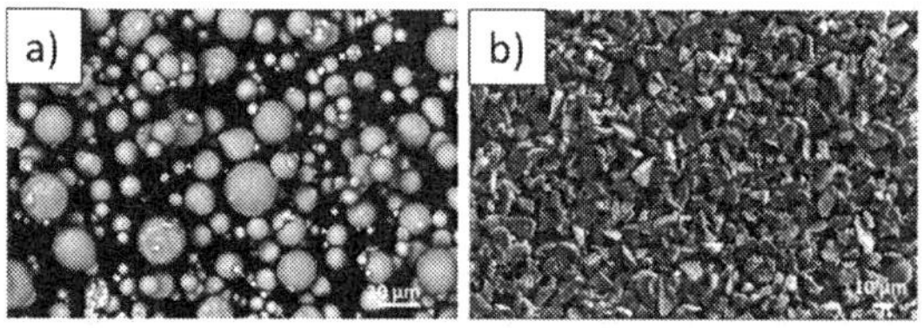

Figure 1. Backscattered electron (BSE) images of a) as-received Al powder and b) as-received SiC powder.

Figure 2. BSE image of the ball milled Al-SiC powders after 6 hours.

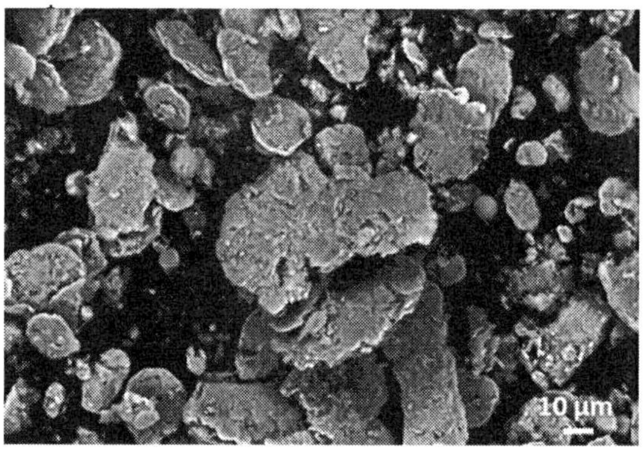

Figure 3. BSE image of the ball milled Al-SiC powders after 12 hours.

Corresponding author: lmiltia@gmail.com

EFFECT OF VC, Cr_3C_2 AND Ru ADDITIVES ON THE MECHANICAL PROPERTIES OF WC-10Co

S.T. Luthuli , L.A. Cornish and D.J. Whitefield

School of Chemical and Metallurgical Engineering, University of the Witwatersrand, and DST-NRF Centre of Excellence in Strong Materials, hosted by University of the Witwatersrand, Johannesburg

Tungsten carbide cobalt, WC-Co, is a composite of tungsten particles with metallic cobalt as a binder, usually made from powders, pressed and sintered at a high temperature[1]. Cemented carbides possess superior mechanical properties and excellent abrasive wear resistance[2], and are commonly used in cutting tools, drilling and wear resistant machine parts. Typical hard-facings on softer materials include WC-Co of varying Co contents. Under abrasive wear conditions, the binder is rapidly removed (being softer), which then allows the hard WC particles to be easily removed[3]. Thus, a technique for strengthening the binder is needed.

To study physical and mechanical properties of WC-Co, samples of different compositions were made. As-received WC, Co, Ru, VC and Cr_3C_2 powders were analysed in a Mastersizer 2000 for particle size, then powders of different compositions, WC-10Co powder mixtures with different additions, were milled in a planetary ball mill with a 4:1 ball-to-powder ratio. The different powder mixtures were then sintered in a spark plasma sintering furnace at 1150°C and 50 MPa for 8 minutes dwell-time. The sintered samples were tested for density in a Sartorius ED224S machine. Specimens were prepared metallographically, by grinding and polishing, and etched with Murakami's reagent. Microstructural analysis was done in a Zeiss FE-SEM with energy dispersive X-ray spectroscopy to study the microstructures and analyse the overall and phase compositions.

The microstructure of the samples are shown in Figs. 1-3. There are limit significant microstructural differences. Fine WC grains were seen in all the samples, as a result of Ru, VC and Cr_3C_2 additives being partially successful as grain growth inhibitors. Figure 2 shows a wide grain size range with many WC/WC contacts without a continuous Co binder. In general, the Co solid solution was not well distributed between the WC grains and as a result Co pools were formed. The VC and Cr_3C_2 dissolved in WC, and Ru dissolved in the binder.

An increase in hardness was observed as a results of the additives introduced to the WC-10Co base material. Hardnesses were calculated to be 1709 ± 36HV$_{30}$ (Fig. 1), 1737 ± 42HV$_{30}$ (Fig. 2) and 1656 ± 35HV$_{30}$ (Fig. 3) compared to the WC-10Co base material with 1564 ± 75HV$_{30}$. The best additive for improved hardness was found to be Cr_2C_3.

References:
1. Upadhyaya, G. (2002) Materials and Design 22(6), 483.
2. Chabretou, V., Alibert, C. and Missiaen, J. (2003) J. Mater. Sci. 38(12), 2581.
3. Siwak, P. and Garbiec, D. (2016) Nonferrous Met. Soc. China, 26(10), 2641.

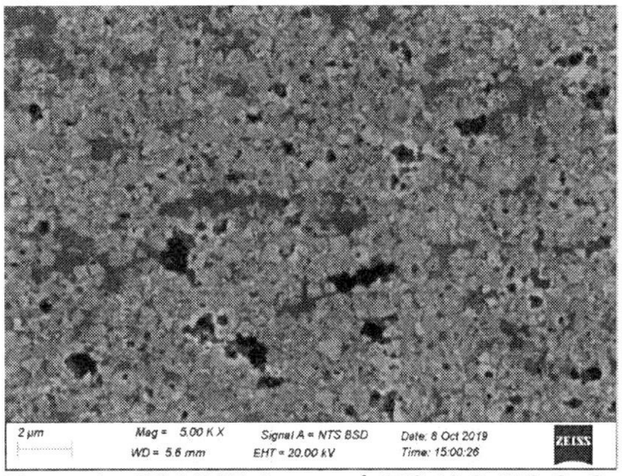

Figure 1. SEM-BSE image of WC-10-0.4VC-Cr_3C_2, showing Co pools (dark contrast) and WC (light).

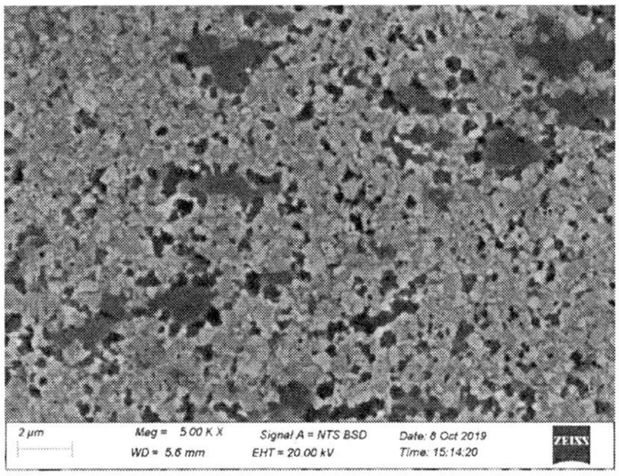

Figure 2. SEM-BSE image of WC-10-0.4Ru-Cr_3C_2, showing WC (light) and Co (dark contrast) with different sized WC grains.

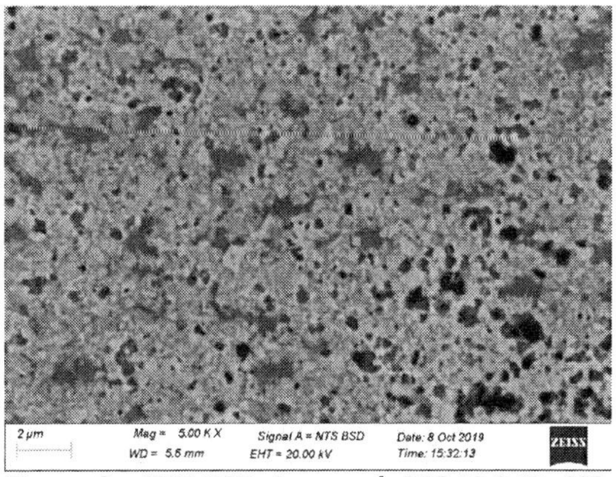

Figure 3. SEM-BSE image of WC-10-0.4Ru-VC, showing WC (light) and Co (dark contrast) with more rounded WC particles.

Corresponding author: 900957@students.wits.ac.za

EFFECTS OF TiC_7N_3 AND WC ADDITIONS ON NbC-BASED INSERTS DURING TURNING OF STRUCTURAL STEEL

R.M. Genga[1,2,3], P. Zeman[4], J. Brajer[4], J. Malý[4], T. Primus[4], M. Pešice[4], S. Huang[5], J. Vleugels[5] and A. Janse van Vuuren[6]

[1]Academic Development Unit, [2]DST-NRF Centre of Excellence in Strong Materials, [3]School of Mechanical, Industrial and Aeronautical Engineering, University of the Witwatersrand, Johannesburg, [4]Research Centre of Manufacturing Technology, Czech Technical University in Prague, Czech Republic, [5]Department of Materials Engineering (MTM), KU Leuven, Belgium, [6]Centre for HRTEM, Nelson Mandela University, Port Elizabeth

Structural steels are used in several applications ranging from construction to transport due to their good combination of tensile strength, durability, corrosion resistance and affordability[1]. In this study, niobium carbide (NbC) based materials were investigated as potential inserts for turning of CSN11109 structural steel because of the comparable hardness of NbC (19.6 GPa)[2] and WC (22.4 GPa)[2], good high temperature properties and chemical stability[2]. Properties such as hardness and abrasion resistance of NbC cemented carbides can be improved through spark plasma sintering (SPS) and additions of secondary hardening phases such as TiC_7N_3 and WC. The NbC based insert performances were compared to WC based inserts.

Cutting inserts were produced from WC-0.8Cr_3C_2-4Mo_2C-4TiC-10Co (M1-S), NbC-4Mo_2C-4TiC-12Ni (wt%) (G1-L) and NbC-14TiC_7N_3-4Mo_2C-20WC-12Ni (wt%) (G3-S) sintered compositions. M1-S and G3-S were prepared by SPS, while G1-L was prepared by pressureless liquid phase sintering (LPS). During dry turning of structural steels, the cutting speed (v_c) was varied between 200-300 m/min, depth of cut (a_p): 0.25-0.50 mm and feedrate: 0.1-0.15 mm/rev. Insert wear was characterised by optical microscopy, scanning electron microscopy (SEM) and annular dark field (ADF) scanning transmission electron microscopy (STEM).

Figure 1 shows the rake faces of the of M1-S, G1-L and G3-S inserts after turning at v_c = 200 m/min and a_p= 0.5 mm. Crater wear was observed in the WC based insert (Fig. 1 (a)) after 16 minutes of machining. Conversely, no crater wear was observed in both NbC based inserts (Figs. 1 (b) and (c)) after 20 minutes of machining. The crater wear was attributed to the dissolution of WC in steel, resulting in atomic dissolution of the insert material in the workpiece chip. Although the M1-S insert had good room temperature hardness (Table 1), poor chemical stability also led to a high flank wear rate (FWR) (Table 1).

The G1-L and G3-L inserts had no crater wear due to NbC's good chemical stability[2]. Although both inserts had good chemical stability, the harder G3-S insert exhibited lower FWR than G1-L (Table 1) due to the improved abrasive resistance from TiC_7N_3 and WC additions, as well as SPS. SPS increased the hardness by refining the microstructure due to lower sintering temperature and shorter dwell time than LPS[2]. The LPS G1-L insert had large NbC grains (~2 µm) and thick binder pools, leading to wear from binder tearing (Fig. 2) and carbide pull-out (attrition). Hence, G3-S insert performed better than the G1-L and M1-S due to its good combination of hardness and abrasion wear resistance from SPS, TiC_7N_3 and WC additions as well as good chemical stability from NbC.

References:
1. Brockenbrough RL and Merritt FS. (1999) Structural Steel Designers Hand Book. USA, McGraw-Hill.
2. Genga R. *et al.* (2018) Int. J. Refract. Met. Hard. Mater. 73, 111.

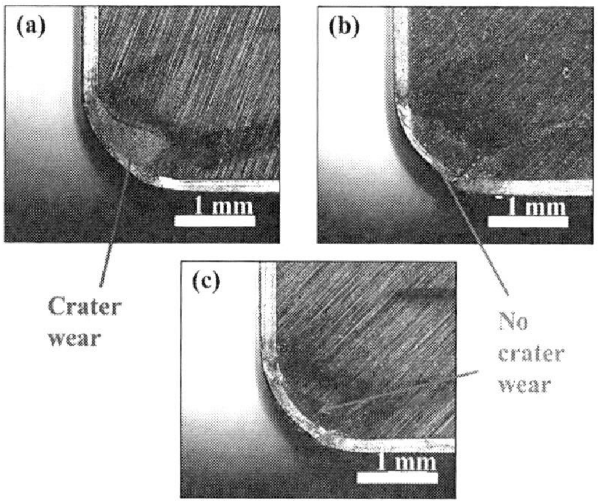

Figure 1. Optical images of the cutting inserts, showing the rake faces of: (a) M1-S, (b) G1-L and (c) G3-S.

Table 1. Mechanical properties and flank wear

Grade	Hardness (GPa)	Flank wear rate (µm/min)
M1-S	15.24 ± 0.18	37.1
G1-L	12.56 ± 0.13	26.3
G3-S	15.28 ± 0.14	21.2

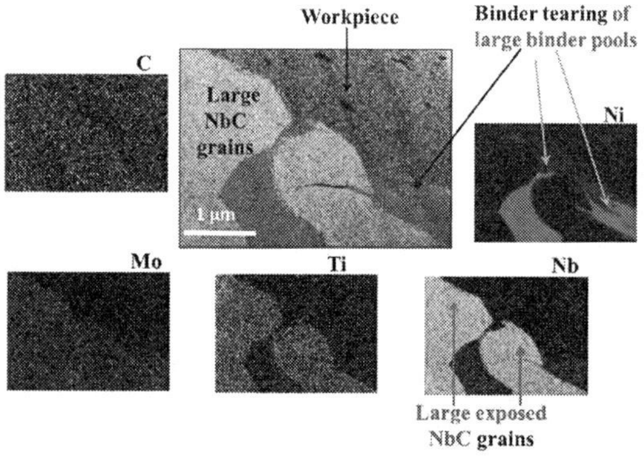

Figure 2. ADF-STEM mapping images of G1-L insert cutting edge/workpiece interfaces, showing: Ni (red), Nb (green), Ti (brown), Mo (purple) and C (yellow)

Corresponding author: Rodney.Genga@wits.ac.za

EVOLUTION OF PRECIPITATES DURING SLAB CASTING, REHEATING AND ROUGHING MILL HOT-ROLLING IN AISI 436 STEEL

A.J. Kruger[1,2] and C.W. Siyasiya[2]

[1]Columbus Stainless (Pty) Ltd, Middelburg, [2]Materials Science and Metallurgical Engineering, University of Pretoria, Pretoria

The automotive industry has continuously been driven by cost cutting and, therefore, there is need for ferritic stainless steels (FSS) to replace the expensive austenitic stainless steels (ASS) without compromising on performance under extreme engineering applications. One of the major short comings of FSS is that the formability properties are significantly lower than ASS. Mo is added to these FSS to improve the high temperature performance and pitting corrosion resistance. A typical steel grade is 436 FSS which contains 0.3-0.5% Nb and 0.8 -1.3% Mo and is not Ti stabilised. With this substantial amount of alloying elements in the steel, it is imperative to understand the precipitation behaviour during casting, slab reheating and roughing mill rolling processes in pursuit of the optimized ridging and formability properties of the final product. In this work, the cast structure and precipitation behaviour of this steel grade have been investigated using the Gleeble 1500^{TM} machine, optical microscope (OM) and scanning electron microscope (SEM) with energy dispersive x-ray spectroscopy (EDS) to identify the precipitates. The Thermo-Calc prediction of the phase transformation is shown in Figure 1.

The as-cast slab samples (100 x 70 mm) were all ground and then polished to 1 micron. The polished samples were examined using the SEM, Figure 2. Thereafter the Bereha's tint was used to etch the steel for 20 min. The specimens were examined using the SEM-BSE and it was observed that the platelets intersectioned each other as may be seen in Figure 3.

In as cast condition, it was found that the complex plate-like Laves phase $(FeCrSi)_2(MoNb)$ which co-existed through heterogenous nucleation with $Fe_3Nb_3(C,N)$ carbo-nitrides formed in the chill zone of the slab. The Laves phase exhibited an orientation relation with the matrix, Figure 2 and 3.

After reheating the samples at 1070 and 1150 ° C for 10 minutes, the precipitates did not completely dissolve but some Nb diffused from the matrix into the Laves phase replacing Mo, Fe and Si to various extents depending on slab depth/cast structure. However, the shape of the precipitate remained unchanged. The precipitates remained unchanged even after simulated roughing mill hot-rolling to a total reduction of 86%.

References:
1. Ahn, Y.S. and Song, J.Y. (2011) Metals and Materials International 17(6), 4.
2. Kato, Y. *et al.* (2010) Materials Transactions, 51(9), 1531.
3. Juuti, T. *et al.* (2014) Journal of Alloys and Compounds 616, 7.

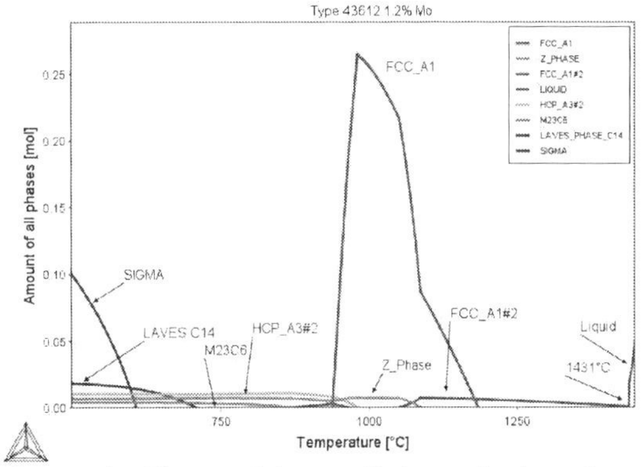

Figure 1. Thermo-Calc prediction of the phase transformation.

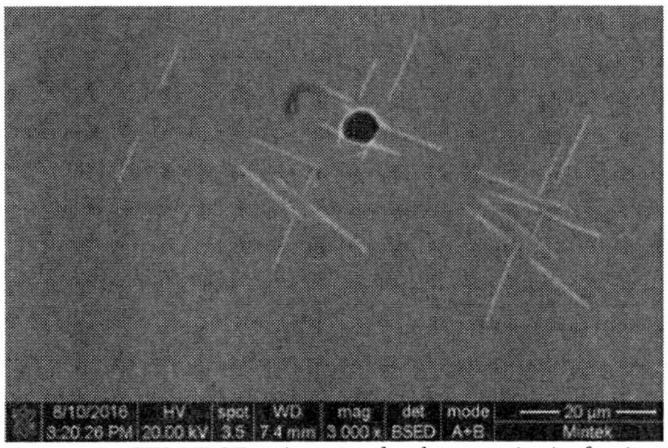

Figure 2. SEM/BSE micrograph, the matrix is ferrite, light grey Laves phase precipitates and the dark phase is a non-metallic inclusion.

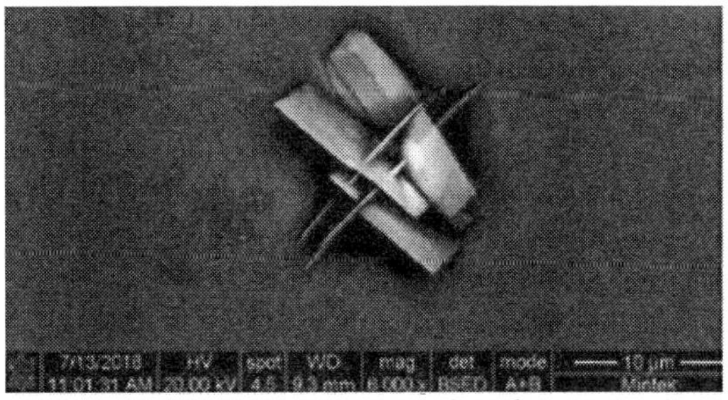

Figure 3. SEM/BSE micrograph of Laves phase plate-like precipitate where the matrix had been etched away

Corresponding author: kruger.jaco@columbus.co.za

EXPERIMENTAL VERIFICATION OF A COMPUTATIONAL APPROACH TO STUDY SHI IMPACT KINETICS

J.H. O'Connell[1], R.A. Rymzhanov[2] and V.A. Skuratov[2]

[1]Centre for HRTEM, Nelson Mandela University, Port Elizabeth, South Africa, [2]FLNR, JINR, Dubna, Russia

A swift heavy ion (SHI, E >1 MeV/amu) loses the largest part of its energy (>95%, 5–50 keV/nm along the ion trajectory) through excitation of the electronic subsystem of the target material. Subsequent relaxation of the excited volume occurs at ultrashort spatial (nanometers) and temporal (femto- to pico-second) scales and cannot be described in the framework of usual macroscopic models and is impossible to follow experimentally. SHI penetration trough different insulators have shown quite different manifestations of structure transformations[1]: amorphous tracks (e.g. $Y_3Fe_5O_{12}$, α-quartz), defected crystalline tracks (e.g. Mg_2AlO_4, Al_2O_3, ZrO_2, TiO_2) or production of isolated point defects and color centers (e.g. MgO, alkali halides). This motivates research aimed at the understanding of mechanisms of track (and surface hillock) formation in these materials.

We compared the responses of single crystalline α-Al_2O_3, MgO and $Y_3Al_5O_{12}$ (YAG) specimens to irradiation with 167 MeV Xe ions at 300 K. Irradiations were performed at the IC-100 cyclotron at FLNR JINR. High resolution transmission electron microscopy (HRTEM) studies were carried out at the Centre for HRTEM using a Cs corrected JEOL ARM-200F TEM operating at 200 kV.

A hybrid simulation approach used in this work consists of two models: Monte Carlo simulation (MC code TREKIS[2-3]) of the electron kinetics, and Molecular Dynamics model of atomic dynamics. TREKIS describes the temporal evolution of excited electrons generated by a SHI as well as the interaction of primary and secondary electrons with the target lattice in an ion track. The resulting distribution of energy transferred to the ionic subsystem of the target is inserted into classical MD code which is used to simulate lattice energy relaxation and further structure transformations in the vicinity of the ion trajectory.

Despite comparable energy deposition into the lattice (especially for radii >1 nm), the passage of 167 MeV Xe produced notably different damaged structures in the investigated oxides. MgO showed no experimentally observable track which is consistent with simulation results. This is due to only a few isolated point defects remaining after relaxation. For Al_2O_3, the remaining damage is similar in nature to that of MgO but at significantly higher concentration. In the case of YAG, the simulated and experimental ABF STEM images have almost perfect correspondence, an almost perfectly circular region of amorphous material with a relatively sharp boundary with no intermediate phase around the edge. Figure 1 shows simulation results of a 167 MeV Xe ion impacting the surface of a YAG crystal (top) and an MgO crystal (bottom). HAADF STEM micrographs (thermally colored to improve contrast) is inset for both systems at the same scale. In the case of YAG, the hillock is amorphous and exhibits a mound-like structure. In contrast, the hillock in MgO has a more spherical shape and is both crystalline and epitaxial to the bulk crystal.

It was found that our combined MC/MD theoretical approach could sufficiently reproduce the observed structures formed in due to 167 MeV Xe impact in MgO, Al_2O_3 and YAG representing non-amorphizable and amorphizable insulators. Good agreement was found between the structures produced by simulation in the bulk and on the irradiated surface with those observed experimentally. This suggests that the model effectively reproduces the most important aspects of the relaxation process and could be used to interrogate the material parameters responsible for determining the material response to excitation by a SHI.

References:
1. Komarov, F.F. (2017) Physics-Uspekhi 60, 435.
2. Medvedev, N.A., Rymzhanov, R.A., Volkov, A.E. (2015) J. Phys. D. Appl. Phys. 48, 355303.
3. Rymzhanov, R.A., Medvedev, N.A. and Volkov, A.E. (2016) Nucl. Inst. Meth. B388, 41.

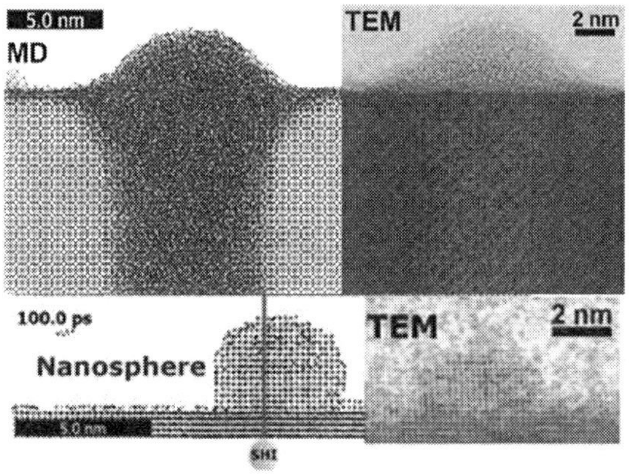

Figure 1. Snapshots of modeled 167 MeV Xe tracks at the surface of amorphizable (YAG) and non-amorphizable (MgO) insulators during mass ejection and after relaxation. The thermal colored TEM micrograph insets are at the same scale as that of the simulation cell.

Corresponding author: Jacques.oconnell@gmail.com

EXTRACTION OF NICKEL OVER-BURDEN ORE USING A FALCON CONCENTRATOR

H.H. Ramashau , K.M. Raselabe and C.K. Thubakgale

Department of Metallurgical Engineering, Vaal University of Technology, Vanderbijlpark, South Africa

Nickel sulphide ores are the main source of produced nickel and these ores require additional capital input on production due to deeper mining requirements and decline in the global reserves [1]. Increasing demand of nickel has pressed for the rapid increase in processing the complex low-grade nickel bearing ores such as overburden and oxide-ore material. Due to its outstanding corrosion resistant characteristics, nickel is a significant metal in the manufacturing of stainless-steel and superalloys[2].

The Nickel oxide ore is very complex to process such actinolite, hence there's no literature available on its processing using falcon concentrator. Falcon concentrator is an enhanced gravity separator method used for concentrating fine/ultrafine minerals[3]. This work aims at investigating the optimum parameter, whereby nickel can be extracted from the overburden material.

Nickel overburden material of 16 kg was reduced into 2 cm inside the Jaw crusher and characterised under an optical microscope to view the thin section of the ore. Further reduction was accomplished, and milling was done at 80% passing with -75 µm. The -75 µm sample was characterised using Atomic Adsorption Spectrometer (AAS) and X-ray diffractometer (XRD), the moisture and density were determined. The sample was then fed in the falcon concentrator with the gravitational force constant throughout with only variation of percentage solids (10 %, 20 % and 35 %) and volumetric flow rate (10, 20 and 30 L/min).

The results showed that the overburden material after XRD contained actinolite constituent of (35 %), Clinochlore (31.3 %), Talc (17%) and magnetite (13.9%). Clinochlore, as a nickel-bearing mineral, confirms its presence in the overburden ore. The presence of Clinochlore associated with actinolite indicates that the deposit is a low-grade metamorphic rock.

The polar images of the digital optical microscopy which were taken under a crossed polarized light. The crossed polarised were taken at 50X and later 200X magnification, as shown in Fig 1. A blue-ish pleochroism suggests the presence of actinolite, with small amounts of talc included. One hundred twenty cleavage lines were observed in the darker region due to thickness.

The dark colour mineral surrounding the rich blue minerals is magnetite (Fig 2) with the softer powdered clear-like powdered version of the blue mineral being chlorite. This is confirmed by the display of internal reflectance (rainbow-like display in the matrix).

The AAS analysis revealed that the sample contains 0.3 g/t nickel concentration together with iron concentration of 4.0 g/t. This implies that the overburden has a very low nickel grade. The nickel content in the overburden ore is very low. From the analysis, it was observed that nickel is hosted in the silicate phase not oxide phase.

References:
1. Elias, M. (2002) CODES Special Publication, 4, 205.
2. Norgate, T. and Jahanshahi, S. (2010) Minerals Engineering 23(2), 65.
3. Oruç, F., Özgen, S. and Sabah, E. (2010) Fuel, 89(9), 2433.

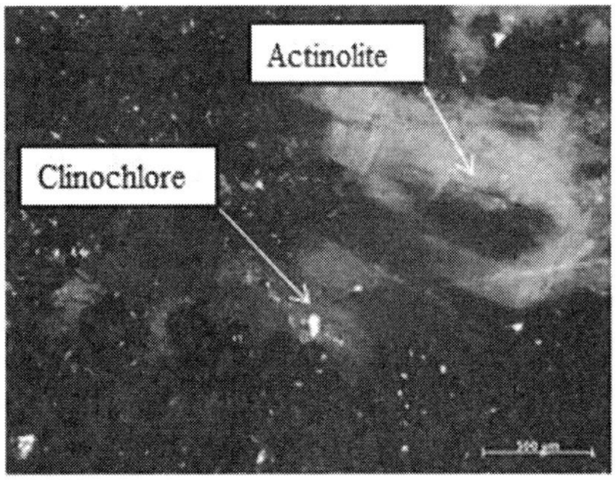

Figure 1. Cross polarized image of overburden ore

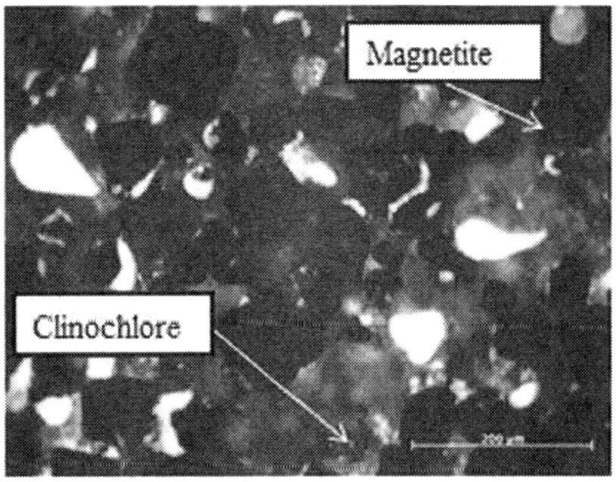

Figure 2. Cross polarized image of overburden ore

Corresponding author: hulisani4m@gmail.com

HELIUM ASSISTED MIGRATION OF SILVER IMPLANTED INTO SiC

C.E. Maepa and T.T. Hlatshwayo

Department of Physics, University of Pretoria, Pretoria

In modern nuclear reactors, safety is improved by adding coating layers. In the pebble bed modular reactors (PBMR), the fuel kernel is coated by four chemical vapour deposited layers[1]. In the coated particle the kernel is encapsulated by chemical vapour deposited silicon carbide (SiC) and three pyrolytic carbon (PyC) layers. Of these layers, the SiC layer acts as the main barrier to fission products (FP) release. These coated particles retain most of the radiological important FPs with the exception of silver[2]. In nuclear environments, SiC is continuously subjected to different irradiations at high temperature in the presence of helium[3]. This leads to the formation of He bubbles in SiC[4]. These bubbles might influence the containment of fission products in SiC. In this study the effect of He bubbles in the migration of silver implanted into SiC is investigated.

Polycrystalline SiC wafers from Cree were used in this study. Ag ions of 360 keV were implanted into SiC to a fluence of 2×10^{16} cm^{-2} at room temperature. A number of as-implanted and virgin samples were implanted with He ions of 17 keV to a fluence of 1×10^{17} cm^{-2} at room temperature. The co-implanted samples were then annealed at 1100°C for 5 hours. The implanted and annealed samples were characterized by transmission electron microscopy (TEM). TEM lamellas were prepared by focussed ion beam (FIB) technique. The migration of implanted silver was monitored by Rutherford backscattering spectrometry (RBS).

TEM micrographs and corresponding selected area diffraction (SAED) patterns of He implanted and He and Ag co-implanted are shown in (Fig. 1). He implantation caused the formation of He bubbles at about 100 nm from the surface-Fig. 1(a). He bubbles are also formed in the co-implanted samples in the same depth as of implanted Ag of about 100 nm. The SAED patterns of the He bubbles is presented by Fig 1(a') and (b') respectively, showing the amorphous diffuse concentric rings. This is contrasted by the SAED pattern observed for the crystalline layer of the un-implanted SiC structure. The formation of He bubbles has been reported in SiC for the same implantation He fluence[4].

The RBS results in (Fig. 2) shows the co-implanted SiC annealed at 1100 °C. The silver as-implanted SiC result is included for comparison. The Arrows in (Fig.1) indicate the surface positions of elements. Co-implantation resulted in more carbon on the surface which resulted in the shift in the high energy Si edge. These were accompanied by implanted Ag shifting towards the bulk. These might be due to formation of He bubbles resulting in the destruction of SiC with Si sublimating leaving C on the surface. Annealing caused the implanted Ag to move towards the surface resulting in some loss of Ag. No migration of implanted Ag into SiC at RT after annealing at 1100 °C for 5 h has been observed[5]. Hence, He bubbles assisted the migration of implanted Ag. More investigations are underway to get more insight in the role of He bubbles in the migration of implanted silver.

References:
1. Van der Berg, N.G. *et al.* (2010) S. Interface Anal. 42, 1156.
2. Dwaraknath, S. and Was, G. (2016) J. Nuclear Mat, 474, 76.
3. Heinisch, H.L. *et al.* (2004) J. Nuclear Mat. 327, 175.
4. Beaufort, M-F. *et al.* (2015) J. Applied Phys. 118, 205904.
5. Abdelbagi, H.A.A. *et al.* (2019) N. Instrum. Methods Phys. Res. B, 451, 113.

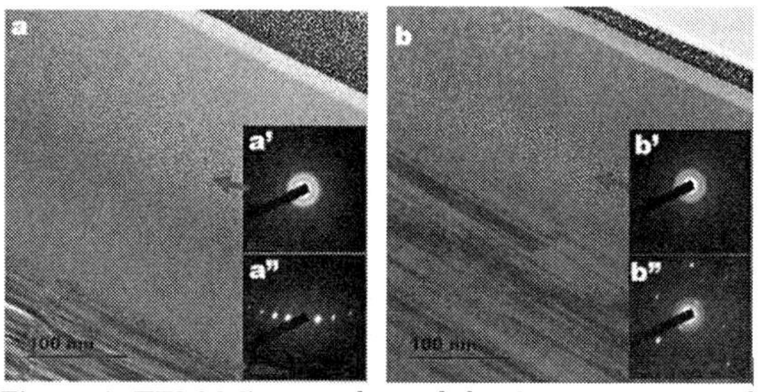

Figure 1. TEM Micrographs and the SAED patterns of the SiC layer with the He implanted (a, a' and a") and He and Ag co implanted (b, b' and b")

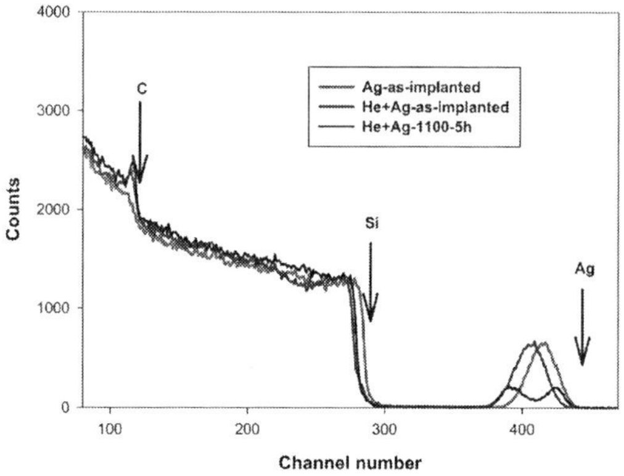

Figure 2. RBS spectra of He and Ag co-implanted into SiC then annealed at 1100 °C for 5h. Ag implanted into SiC spectrum is included for comparison.

Corresponding author: charity.maepa@up.ac.za

IN SITU OBSERVATION OF DEFECT FORMATION IN SHI INDUCED TETRAGONAL TRACKS IN MONOCLINIC ZIRCONIA

M.E Lee[1], J.H. O'Connel[1] and V.A. Skuratov[2]

[1]CHRTEM, Nelson Mandela University, Port Elizabeth, South Africa, [2]FLNR, JINR, Dubna, Russia

Zirconia (ZrO_2) is a polymorphic oxide that exists in three different crystal structures below its melting point namely, the high temperature phases cubic and tetragonal as well as the low temperature monoclinic phase[1]. The tetragonal to monoclinic transformation mechanism in bulk materials has received considerable attention and there have been a number of modeling approaches which have been adopted in the interpretation of the phase transformation namely the thermodynamic based and crystallographic based models and more recently the phase field model[2]. Monoclinic to tetragonal transformation has been reported for swift heavy ion (SHI) irradiated monoclinic single and polycrystalline bulk zirconia but only at fluences high enough for track overlap[3]. The observation of single tetragonal tracks in monoclinic zirconia was recently first reported for low Xe ion fluences by Lee and co-workers[4]. In this work, in situ TEM analysis is reported for the formation of defect loops along tetragonal tracks produced by swift heavy ions during heating.

A monoclinic zirconia single crystal was irradiated with 167 MeV Xe ions at the FLNR cyclotron complex in Dubna, Russia to a fluence of 2×10^{10} cm^{-2}. Irradiation was performed at room temperature and TEM lamellae were prepared in cross section and plan view geometry using an FEI Helios Nanolab FIB. These lamellae were examined using a Dens in situ heating holder in a double Cs corrected ARM 200F operating at 200 kV in both TEM and STEM mode. In situ heating of the TEM lamella was carried out in the range 100 – 400 °C.

Figure 1 shows a BF TEM micrograph of a single tetragonal ion track in monoclinic zirconia. The track geometry was observed to be rectangular with a width of approximately 2.5 nm. However, the tetragonal tracks are are shown to be discontinuous, forming segments which have an average length of about 30 nm. In addition, the discontinuous tetragonal segments are misaligned by approximately 9 degrees relative to each other. This misalignment is explained in terms of the angular mismatch between the tetragonal and monoclinic unit cells along the c-axis. The dark contrast observed at the discontinuities is an indication of the large strain energy due to the crystalline mismatch. Figure 2 shows a BF TEM micrograph of the defect loops formed after annealing in the range 200 - 400 °C along the tracks which have transformed to the monoclinic structure. The position of the defect loops after annealing are found to coincide with the discontinuities for the tetragonal segments. The small size (5-10 nm) of these loops made it difficult to determine their exact crystallographic nature. It is therefore concluded that oxygen possibly diffused to the strained areas at the tetragonal discontinuities during heating to form oxygen defect loops, which is consistent with the theory that oxygen diffusivity increases under stress conditions[5].

References:

1. Lumpkin, G.R. (1999) J. Nucl. Mater. 274, 206.
2. Mamivand, M. *et al.* (2013) Acta Mater. 61, 5223.
3. Degueldre, D. and Hellwig, Ch. (2003) J. Nucl. Mater. 320, 96.
4. Lee, M.E., O'Connell, J.H. and Goosen, W.E. (2015) Proc. Microsc. Soc. South. Afr. 45, 43.
5. Kushima, A. and Yildiz, B. (2010) J. Mater. Chem. 20, 4809.

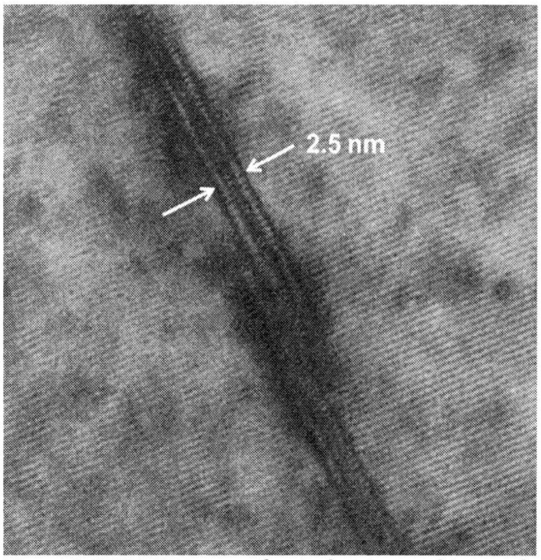

Figure 1. TEM BF image of a single tetragonal ion track in monoclinic zirconia

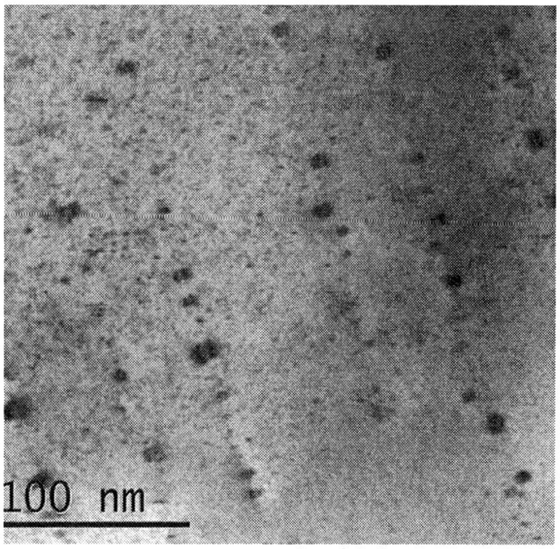

Figure 2. BF TEM micrographs of defect loops formed along ion tracks after in situ heating.

Corresponding author: michael.lee@mandela.ac.za

INCORPORATION OF TRANSITIONAL AND NOBEL METALS NANOSTRUCTURES IN P3HT:TiO_2 BLENDS

P.J. Maake[1,2], E.I. Iwuoha[2] and D.E. Motaung[3]

[1]DST/CSIR National Centre for Nanostructured Materials, Council for Scientific Industrial Research, Pretoria, [2]SensorLab, Department of Chemistry, University of the Western Cape, Cape Town, [3]Department of Physics, University of Limpopo, Polokwane

The development of new nanomaterials with multifunctional properties that can be utilized for solar harvesting and gas sensing with improved light absorption, improved charge transport performance, enhanced gas sensitivity and selectivity indicates one of the leading scientific challenges driven by enormous energy demands and emission of toxic gases in industrial processes, ensuing hazards in public health, industries and national security. Thus, herein, we focus on the synthesis of TiO_2 doped with transitional (Cu) and noble metals (Ag and Au) and their incorporation in P3HT polymer matrix for inorganic-organic photovoltaic solar cells and low operating temperature chemiresistive sensors. Synthesis of TiO_2 mixed with different concentration of Au, Ag and Cu were prepared using hydrothermal method with the objective of defining the structural and optical properties to probe their influences on the polymer solar cell devices.

The presence and the role of the metallic dopants in the hybrid nanostructure (P3HT:TiO_2-Ag, P3HT:TiO_2-Au, P3HT:TiO_2-Cu, P3HT:TiO2-Cu-Ag and P3HT:TiO_2-Ag-Au) were characterized in detail using X-ray diffraction (XRD), scanning electron microscopy-energy dispersive X-ray (SEM-EDS) study, transmission electron microscopy (TEM) analyses. The low and high resolution TEM analyses in Figure 1 and Figure 2 demonstrate that Au nanoparticles are homogenously distributed across the TiO_2 matrix. TEM image display an interlayer spacing of 0.350 nm which correspond to (101) lattice planes of anatase TiO_2, Wherefore the d-spacing of 0.233 nm match the (111) lattice planes of Au nanoparticles, which is consistent with the results reported by Zhu *et al.*[1] and Naik *et al.*[2]. the XRD patterns shows no impurity peaks revealing dopants had inserted TiO_2 crystal lattice.

The surface plasmonic resonance behaviour and charge transfer, owing to the delocalization of photoexcited electrons from the conduction domain of polymer to the domain of metallic dopant was probed using electron spin resonance, ultra-violet visible absorption, photoluminescence and x-ray photoelectron spectroscopy measurements. The role of transitional and noble metals on the improved gas selectivity, sensitivity, response-recovery times, limit of detection and long-term stability were also investigated towards various target gases at low operating temperatures.

I acknowledge the Department of Science and Technology (DST) and Council for Scientific and Industrial Research (CSIR) for funding the work.

References:
1. Zhu, S. *et al.* (2012) Applied Catalysis B: Environmental 119, 146.
2. Naik, G.K. *et al.* (2019) Journal of Alloys and Compounds 771, 505.

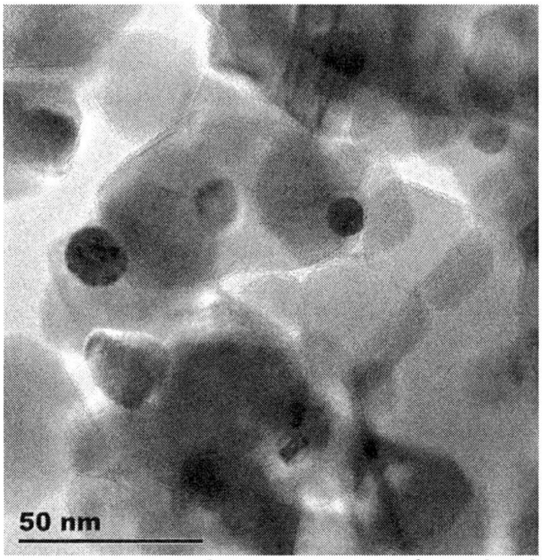

Figure 1. Low resolution TEM image of Au doped TiO_2

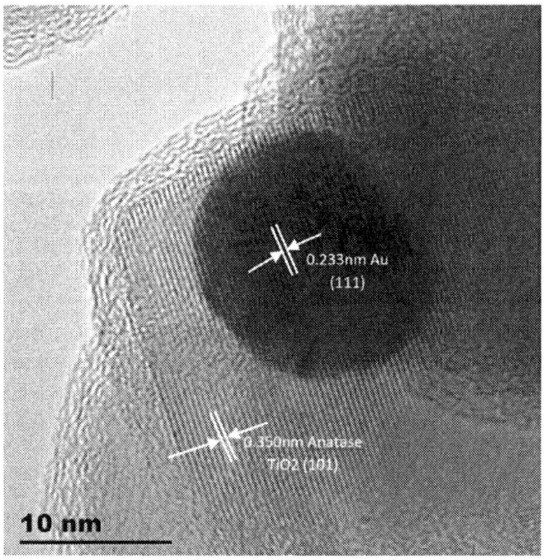

Figure 2. High resolution TEM image of Au dope TiO_2

Corresponding author: pmaake@csir.co.za

INFLUENCE OF OXYGEN ON THE MICROSTRUCTURAL DEVELOPMENT OF Ti-46.5Al-0.8Ta (at.%) ALLOY PRODUCED BY SPS

P.V. Cobbinah[1], W.R. Matizamhuka[1], R. Machaka[2,3], M.B. Shongwe[4] and Y. Yamabe-Mitarai[5]

[1]Dept. of Metallurgical Engineering, VUT, Vanderbijlpark, [2]Advanced Materials and Engineering, Manufacturing Cluster, CSIR, Pretoria, [3]School of Mining, Metallurgy and Chemical Engineering, UJ, Johannesburg, [4]Dept. of Chemical, Metallurgical and Materials Engineering, TUT, Pretoria, [5]Structural Materials Group, National Institute for Materials Science, Tsukuba, Japan

TiAl alloys have received increasing attention over the years and are envisaged as potential replacements of nickel-based superalloys (NBSAs) owing to combination properties such as low density (3.9-4.2 g.cm^{-3}), high strength to weight ratio (up to 1000 MPa) and substantial resistance to corrosion and oxidation[1]. However, at room temperature (RT) TiAl alloys exhibit low ductility, which makes their fabrication difficult[1]. The use of the powder metallurgy (PM) processing route, rapid solidification such as spark plasma sintering (SPS) and microalloying help improve the RT ductility of TiAl alloys.

The microstructural development of 0.8 (at.%) tantalum (Ta) added to Ti-46.5Al alloy produced using PM and SPS techniques was studied. Elemental powders of Al and Ti were milled separately with stearic acid (6 hours) and in ethanol (4 hours). A ball-to-powder ratio (BPR) of 5:1 was used for both millings. The weighed alloy composition was placed in a Turbula shaker-mixer for 4 hours and mechanically alloyed at a speed of 350 rpm and BPR of 10:1 using a Retsch PM 100. The SPS process was carried out in a vacuum at a heating rate of 473 K.min^{-1}, applied pressure of 32 MPa and dwell time of 10 mins at 1423 K. A Jeol JSM-7000F scanning electron microscope (SEM) equipped with an energy dispersive spectroscope (EDS) was used for the microstructural analysis.

The SEM analysis shows the microstructure of the alloy was inhomogeneous. The predominant phases were the α_2-Ti_3Al, γ-TiAl and α-Al_2O_3 phases. In Fig 1. the dark grey contrast constituting the majority of the microstructure was the γ-TiAl phase. The light grey contrast formed the duplex $\alpha_2+\gamma$ phase[1]. The presence of high oxygen content in the alloy led to the precipitation of numerous dark phases analysed to be α-Al_2O_3. The inhomogeneous microstructure could be attributed to the high oxygen content in the TiAl alloy. High oxygen presence during solidification tends to expand the α-phase field and raise the eutectoid temperature of Ti-Al alloy systems[2]. Hence during cooling, oxygen favoured the chemical ordering of $\alpha \rightarrow \alpha_2$ as a result of the significant difference in solubility of oxygen between the γ and α_2, with the equiaxed γ phase unaltered. Figure 2 shows three different sizes of equiaxed grains noticed at certain parts of the alloy.

The presence of oxygen regarded as a α-stabilising element is known to dissolve preferentially in the α-phase[2]. Therefore, when the ratio of Ti/Al is <1 or Ti/Al ratio $\geq$1 with high oxygen content, precipitation of α-Al_2O_3 is likely to occur in the α-phase then in the α_2 and γ-phases[3] as seen in Fig. 1.

The presence of oxygen and the slow diffusivity of Ta in TiAl alloys influenced the diffusion, chemistry and thermodynamic processes taking place in the alloys' microstructure formation[1].

References:
1. Appel, F., Paul, J.D.H. and Oehring, M. (2011) Gamma Titanium Aluminide Alloys: Science and Technology. John Wiley & Sons.
2. Naka, S. (1996) Curr Opin Solid St Mat. 1(3), 333.
3. Kawabata, T., Abumiya, T. and Izumi, O. (1992) Acta Metal. et Mat. 40(10), 2557.

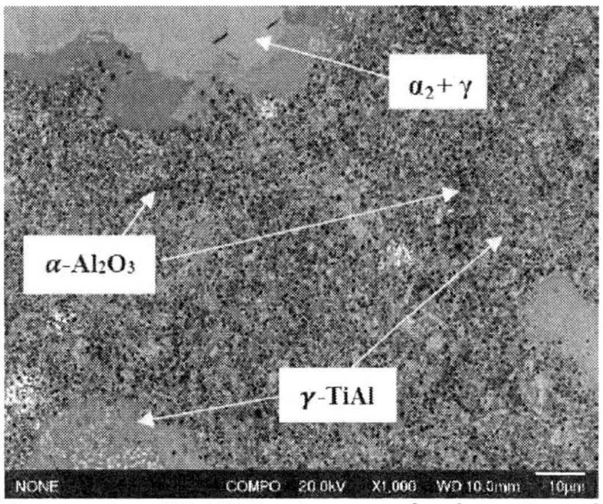

Figure 1. SEM-BSE image of the sintered Ti-46.5Al-0.8Ta (at.%) alloy.

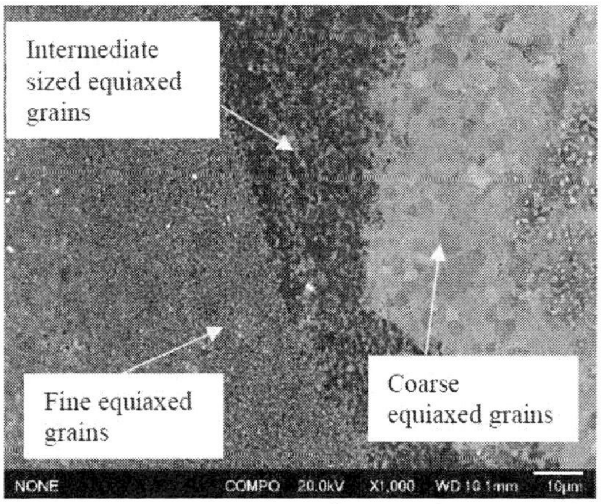

Figure 2. SEM-BSE image of equiaxed grains in the sintered Ti-46.5Al-0.8Ta (at.%) alloy.

Corresponding author: prinzcobbs@gmail.com

INTERFACIAL MICROSTRUCTURE OF WC-6wt.%Co TORCH BRAZED JOINTS USING Ag-BASED FILLER ALLOY

N.P. Mphasha[1,2] and D.J. Whitefield[1,2]

[1]School of Chemical and Metallurgical Engineering, University of the Witwatersrand, Johannesburg, [2]DST-NRF Centre of Excellence in Strong Materials, hosted by University of the Witwatersrand, Johannesburg

In recent years, brazing has received great attention for joining cemented carbides together or to a base material due to its simplicity, high joint strength, good repeatability and low cost[1,2]. Jiang et al.[3] and Hasanabadi et al.[4] obtained sound WC-Co/carbon steel furnace brazed joints using Ag-based filler alloys. Hence, there is potential of brazing at a common interface.

In order to study the effect of filler alloy thickness on the interfacial microstructural evolution, WC-6wt.%Co hardmetals were brazed at a common interface using a Ag-Cu-Zn-Mn-Ni filler alloy. The samples were torch braze under atmospheric conditions at 710 ºC, held for 30 seconds and then cooled in air. Brazed samples were then sectioned across the joint, mounted, ground and polished. Samples were analysed using a Zeiss SEM equipped with energy dispersive X-ray spectroscopy to identify the microstructures and to determine the phase compositions. XRD was conducted using the Bruker D2.

Cross-sectional morphologies of the brazed joints show that metallurgical bonding between the cemented tungsten carbides and the filler alloy were achieved without any voids or cracks. Dark island-like α-Cu (s.s) phases nucleated along the WC-6wt.%Co/filler alloy interfaces (Fig. 1), suggesting the occurrence of metallurgical reaction and diffusion of metallic elements during the brazing process. In addition to the the α-Cu phases, the brazed joint is composed of a light α-Ag (s.s) phase. The micrographs suggest that the nucleation of the α-Cu phases started from the interface towards the center of the interlayer.

Increasing the filler alloy thickness to 18 μm (Fig. 2) promoted the diffusion of metallic elements across the WC-6wt.%Co/filler alloy interface, and decreased the formation of the α-Cu (s.s) phases. Hence, the formation of a leaf-like Ag-Cu-Zn eutectic phase which was confirmed by XRD. From EDS line-scan analysis, across the interfaces, there is a tendency of the α-Cu (s.s) phases rich in Mn and Ni migrating towards the interface due to the good solubility of Ni in Co[4]. This suggested good wettability of the WC-Co surfaces by the filler alloy. Further, the EDS analysis shows that Co diffused from WC-Co base samples into the interlayer. Future research work will include studying the effect of longer holding times and post-braze heat treatment on the properties of the brazed joints.

References:
1. Jacobson, D.M. and Humpston, G. (2005) Principals of Brazing, Ohio, ASM International Materials Park.
2. Saeed, A.M. et al. (2010) Mater. Des. 31(7), 3339.
3. Jiang, C. et al. (2016) Mater. Process. Technol. 229, 562.
4. Hasanabadi, M. et al. (2017) Trans. Nonferrous Met. Soc. China 27, 2638.

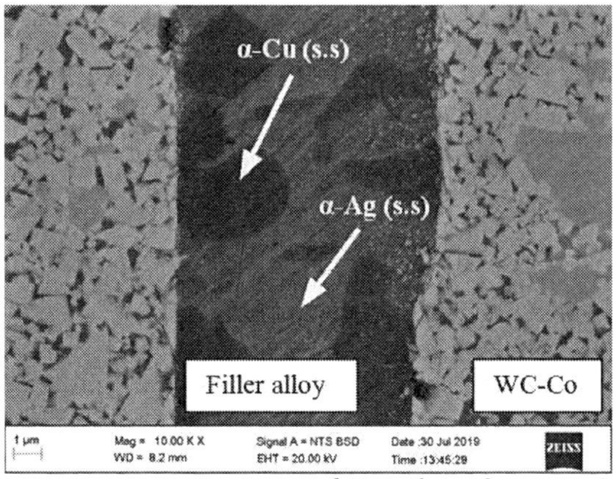

Figure 1. SEM-BSE image of joints brazed at 710 °C for 30 sec with filler alloy of 9 μm in thickness.

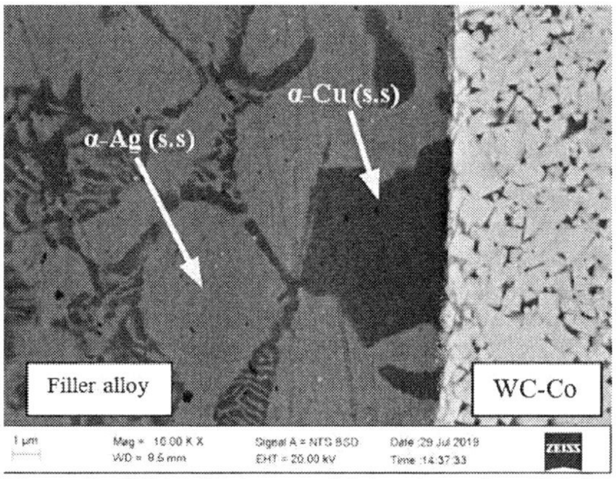

Figure 2. SEM-BSE image of joints brazed at 710 °C for 30 sec with filler alloy of 18 μm in thickness.

Corresponding author: 673306@students.wits.ac.za

K-MEANS CLUSTERING STUDY OF PRECIPATE EVOLUTION IN 12%Cr TMF STEELS DURING LONG-TERM CREEP

W.E. Goosen and J.E. Westraadt

Centre for HRTEM, Nelson Mandela University, Port Elizabeth

Tempered martensite ferritic (TMF) steels with 9-12% Cr are often used for critical components in fossil-fired power plants that operate in the creep range at temperatures up to 650 °C over a lifespan of 200 kh. Precipitate hardening (PH) is regarded as a significant strengthening mechanism to obtain long-term creep strength in 9-12% Cr steels[1]. The Cr-rich $M_{23}C_6$ and V-rich MX carbo-nitrides are the most important precipitates concerning PH. The Mo-rich Laves phase[2] (>200 nm) and the Z-phase[3] (>1 um) that form during long-term creep are considered as microstructure instabilities detrimental to creep-strength. This study investigated the formation and chemical evolution of the precipitate species in a 12% Cr TMF steel during a long-term (139 kh) uniaxial creep test.

The chemical information of precipitates within X20CrMoV12-1 steel as a function of creep-time (virgin condition, 12 kh, 51 kh, 81 kh, 139 kh) was analysed. Samples were cut from a number of interrupted creep specimens (tested under constant load of 120 MPa at 550 °C) and polished using standard polishing methods. Due to the small size of the precipitates as well as the effect of the steel matrix, it was necessary to prepare carbon extraction replicas. Energy-filtered TEM (EFTEM) was used to map the elemental distribution of the precipitates. The samples were then imaged using high angle annular dark-field (HAADF) scanning transmission electron microscopy (STEM) and chemical information obtained using energy dispersive spectroscopy (EDS). K-means clustering[4] was used to assign each measurement to a particular precipitate species.

The EFTEM map in Fig. 1 shows the presence of Cr-rich $M_{23}C_6$ and V-rich MX precipitates inside the virgin steel. The inset shows the atomic% Cr versus V as determined by STEM-EDS. Data points corresponding to the $M_{23}C_6$ and MX precipitates have been coloured green and red, respectively, as determined by K-means clustering. The EFTEM map in Fig. 2 shows the precipitate distribution of the steel that has been exposed to long-term creep loading (139 kh). Diffraction analysis confirmed the formation of the Z-phase (orange). The inset shows the presence of two extra clusters of EDS data points corresponding to Z-phase and the Laves phase (identified using HAADF STEM and EDS).

Quantitative characterisation of the precipitation in TMF steels, especially the transformation of MX into a more thermodynamically stable Z-phase, can provide valuable insight into factors that affect long-term creep-strength. This knowledge can be used to predict the remaining life of power plant components.

References:
1. Hald, J. (2016) T. Indian I. Metals 69(2), 183.
2. Aghajani, A. *et al.* (2009) Scr. Mater. 61, 1068.
3. Danielsen, H.K. and Hald, J. (2009) VGB powertech 5, 68.
4. Pedregosa, F. *et al.* (2011) J. of Mach. Learn. Res. 12, 2825.

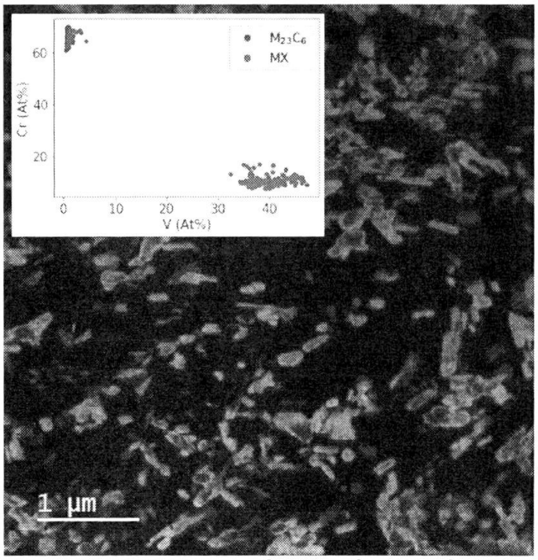

Figure 1. EFTEM map showing the Cr-rich $M_{23}C_6$ (green) and V-rich MX (red) precipitates present in the virgin steel and the corresponding at% V versus Cr scatterplot (see inset).

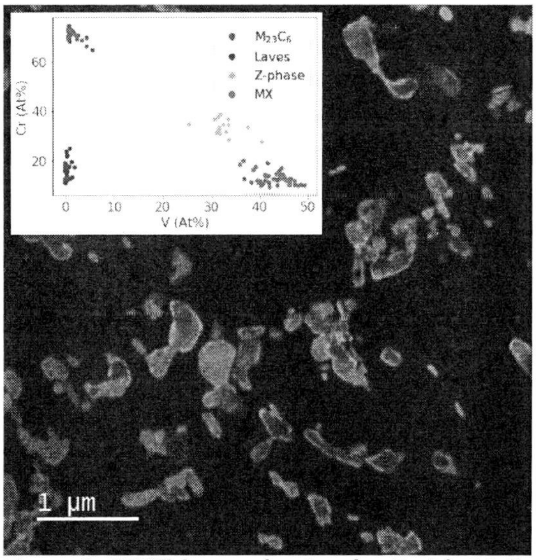

Figure 2. EFTEM map showing the Cr-rich $M_{23}C_6$ (green) and V-rich MX (red) precipitates as well as the presence of Cr-V rich Z-phase (green+red) for the steel exposed to long-term creep. Inset shows the at% V versus Cr scatterplot.

Corresponding author: William.Goosen@mandela.ac.za

LOW PRESSURE VAPOUR PROCESSED HYBRID PEROVSKITE THIN FILMS FOR PHOTOVOLTAIC APPLICATION

S. Ngqoloda[1], C.J. Arendse[1], T.F.G. Muller[1] and C. Oliphant[2]

[1]Department of Physics and Astronomy, University of the Western Cape, Cape Town, [2]Materials Metrology, NMISA, Pretoria

Organic-inorganic hybrid perovskite thin films are promising candidates for photovoltaic technology with lab scale power conversion efficiency (PCE) escalated from 3.8% in 2009 to more than 24% recently[1]. The frequently employed perovskites deposition methods are spin coating from a solution and thermal evaporation of precursor powders[1]. Herein, methyl-ammonium lead iodide ($MAPbI_3$) hybrid perovskite film is deposited by a facile two-step low pressure vapour deposition in a chemical vapour deposition (CVD) reactor.

Lead iodide (PbI_2) thin film was deposited in the first step using a thermal CVD tube furnace. This film was converted into $MAPbI_3$ perovskite using the same CVD chamber in the second step during its exposure to methyl-ammonium iodine (MAI) vapour. The films were deposited on clean glass substrates. The phase composition and the crystal structure of the films were identified by X-ray diffraction (XRD) using Panalytical Empyrean X-ray diffractometer with Cu Kα radiation (1.5406 Å). Surface morphology and cross section were investigated by using a Zeiss Cross Beam 540 Focused Ion Beam Scanning Electron Microscope (FIB-SEM) operated at acceleration voltages of 1-3 kV.

The as deposited PbI_2 film is crystalline with intense XRD peaks as shown by the bottom XRD pattern of Fig. 1a. This film has a preferred (001) orientation and is indexed to a hexagonal (*P-3m1*) structure[2]. The diffraction peaks of (001), (002), (003), and (004) centred at ~12.74°, 25.59°, 38.76°, and 52.51° 2θ, respectively, all belong to PbI_2. The converted $MAPbI_3$ perovskite thin film is also crystalline as shown by the top XRD pattern in Fig. 1a, with strong diffraction peaks at ~14.05°, 19.94°, 28.39°, 31.81°, and 40.61° 2θ, assigned to the (110), (112), (220), (310), and (224) planes, respectively. All the perovskite peaks are indexed to a tetragonal (*I4/mcm*) phase[3]. There are no impurities on the converted perovskite film demonstrating its pure phase and the advantage of using low pressure CVD to prepare perovskite films[1,3].

A flat and compact PbI_2 film is observed on the SEM micrograph in Fig. 1b with average grain size of 598 ± 119 nm, measured from the SEM micrograph. The flat surface of the PbI_2 film is due to a layer-by-layer growth of repeating I-Pb-I monolayers of PbI_2 covalently bonded and separated by a weak van der Waals bond[2]. These layers are stacked along the c-axis, perpendicular to the substrate, and growing along the [001] direction as confirmed by the XRD pattern. This layer-by-layer growth is favourable for conversion to perovskite due to an easy diffusion of MAI molecules within the PbI_2 network. The grains of PbI_2 are hexagonal platelets coalesce to form a compact film; some of these platelets are clearly visible on the surface of the film as seen in Fig. 1b.

The converted $MAPbI_3$ perovskite film during the second step had larger, well defined grains with an average grain size of 886 ± 226 nm, even though there are grains larger than a micron in size as shown in Fig. 1c. Larger grains of perovskites are beneficial for good solar cell device performance due to reduced trap states at grain boundaries. The increase in grain size is related to lattice volume expansion from 0.123 nm^3 of as-deposited PbI_2 to 0.979 nm^3 of tetragonal $MAPbI_3$ perovskite during the conversion step[3]. Cross sectional SEM of the perovskite is shown in Fig. 1d with uniform thickness of about 450 nm and grains extending from film surface to film-substrate interface.

The as-grown PbI_2 crystalline film with flat and compact grains was successfully converted into a homogeneous and pin-hole free $MAPbI_3$ perovskite film with larger grains. The uniform perovskite film thickness is an advantage for device up-scaling.

References:
1. Bulovic, V. *et al.* (2019) Energy Environ. Mater. 2, 119
2. Sun, H. *et al.* (2014) J. Cryst. Growth. 405, 29
3. Baikie, T. *et al.* (2013) J. Mater. Chem. A, 1, 5628

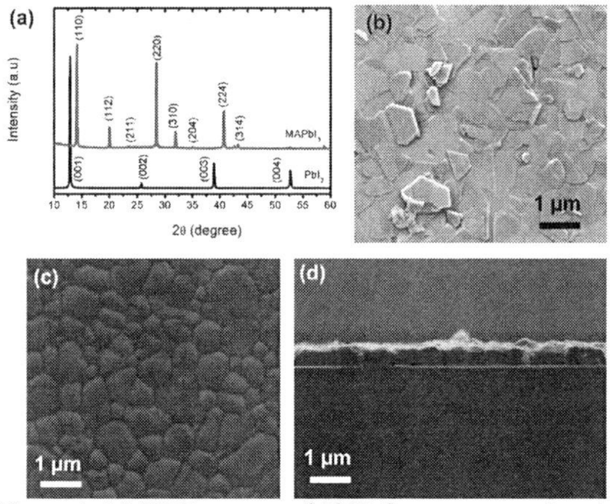

Figure 1. (a) XRD pattern of PbI2 (bottom pattern) and converted MAPbI3 perovskite (top pattern), (b) and (c) are PbI2 and MAPbI3 planar SEM micrographs, respectively, (d) is the perovskite cross sectional SEM micrograph. $_2$ (bottom pattern) and converted $MAPbI_3$ perovskite (top pattern), (b) and (c) are PbI_2 and $MAPbI_3$ planar SEM micrographs, respectively, (d) is the perovskite cross sectional SEM micrograph.

Corresponding author: 2807183@myuwc.ac.za

LOW-TEMPERATURE MARTENSITE DECOMPOSITION IN LPBF-PRODUCED TI6AL4V

G.M. Ter Haar and T.H. Becker

Materials Engineering group, Dep. of Mechanical & Mechatronic Engineering, Stellenbosch University, Stellenbosch

Laser powder bed fusion (LPBF) is a novel powder metallurgy manufacturing technique in which a net-shaped part is additively manufacturing through sequentially melting pre-alloyed metal powder in a layer-wise manner using a high-power laser. LPBF has been successfully applied to Ti6Al4V. This alloy is extremely popular in the aerospace and medical industries due to its high strength-to-weight ratio and biocompatibility. The process and manufacturing chains of LPBF are still in its primary stages, and therefore the achievable mechanical properties of the LPBF-produced Ti6Al4V parts are not optimal compared to its wrought produced counterpart. Poor part ductility is of major concern and is largely attributed to the meta-stable martensitic microstructure[1] (α'). This study aims to develop a deeper understanding of α' decomposition at temperatures below 650 °C to contribute to the body of researched aimed to improve the material's mechanical properties. The results of the study are directly applicable to in-situ base plate heating techniques and the potential influence of parts size on microstructure.

LPBF samples were manufacture using an EOS M280 machine with default process parameters. Heat treatments were conducted using a box furnace at four temperatures: 427, 480, 560 and 610 °C. Holding times varied from 5 minutes to 30 hours. Micro-indentations were carried out using an Emcotest DuraScan automatic indentation machine. A diamond-shaped indenter was used, and measurements were carried out according to ASTM standard E348-10. Load application was 2 kilograms force for 10 seconds. Scanning electron microscopy (SEM) and micro-indentation samples were slow-cut using a diamond-tip cutting blade thereafter ground and polished using recommended Buehler metallography procedures and equipment. Samples prepared for Scanning Transmission Electron Microscopy (STEM) were cut using fine ion beam (FIB) milling in a FEI-HELIOS-Nanolab-650 at the Centre for High Resolution Transmission Electron Microscopy (CHRTEM) at Nelson Mandela University. A Zeiss MERLIN FE-SEM at the Centre for Analytics Facilities (CAF) at the University of Stellenbosch was used in conjunction with a backscatter detector (BSD) to monitor phase transformation. To identify high resolution microstructural transformation, STEM analysis at the CHRTEM was done using a JEM-ARM200F microscope. Electron energy loss spectroscopy (EELS) was used in conjunction with STEM to map elemental composition also at the CHRTEM.

Vickers micro-indentation results reveal an increase in material hardness from an as-built hardness (345 HV2) of all heat-treated samples. Material hardness peaked after only 1 hour for all samples with the highest hardness being achieved (412 HV2) by the sample annealed at 480°C. Material softened after 30 hours with the sample annealed at 610 °C showing the larges decrease in material hardness. Microscopy analysis revealed the nucleation of fine vanadium-rich precipitates at α' grain boundaries and dislocations. Although too fine to quantify, fluctuations in material density was also observed inside grains. With an increase in temperature and hold time, these precipitates agglomerated and grew along α grain boundaries. Since β phase is stabilised by vanadium, these precipitates can be confirmed as β phase or a pre-β phase.

Evidence of martensite decomposition occurring through nucleation-and-growth is clear. The unanticipated material embrittlement is likely caused by α/β grain boundary strengthening. Although β phase is softer than α phase, the high number of α/β grain interfaces and small grain size of β phase caused boundary strengthening to be the dominating effect during early stages of phase transformation. Further growth of the β phase with time and temperature was shown to increase the effect of β deformation vs α/β interface strengthening thereby softening the material. This study contributes to the limited understanding of the initial stages of martensite decomposition at low temperatures.

References:
1. Ter Haar, G.M., Becker, TH. (2018) Materials. 11(1), 146.

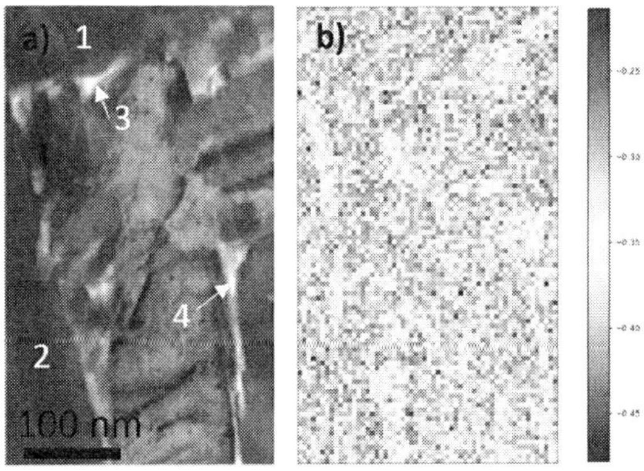

	Weight % (σ)		
Spot	Ti	Al	V
1	89.4 (0.5)	4.3 (0.3)	4.3 (0.4)
2	89.5 (0.2)	4.3 (0.1)	6.2 (0.2)
3	81.9 (0.2)	4.5 (0.1)	13.6 (0.1)
4	78.5 (0.2)	3.8 (0.2)	17.8 (0.1)

Figure 1. STEM analysis (a) HAADF-STEM image (b) EELS map of vanadium, (c) quantification of selected points in the HAADF-STEM image

Corresponding author: gterhaar@sun.ac.za

METAL POWDER SPHEROIDISATION USING RADIO-FREQUENCY THERMAL PLASMA

R. van der Merwe[1,2], H. Bissett[1,2] and E. Snyders[1,2]

[1]Applied Chemistry Department, South African Nuclear Energy Corporation SOC Limited (Necsa), Pelindaba, [2]The Advanced Materials Initiative (AMI) of the South African Department of Science and Technology (DST), Pretoria

Additive manufacturing (AM) requires dense spherical feed powders[1] of a narrow particle size distribution with good flow properties as this affects the packing density and sintering mechanism of the manufactured parts. In-flight spheroidisation by thermal plasma[2] presents an attractive method to produce high melting point metals for AM.

The capabilities of the 15 kW TEKNA radio frequency plasma system was evaluated by spheroidisation experiments in a Ar/H_2 plasma at 13 kW plasma plate power feeding irregular shape titanium (Ti) metal powder (size: 100 – 80 mesh) at a rate of 0.22 kg/h. At the given plasma plate power, 90 % of the particles were spheroidised resulting in improved powder flow characteristics and a density increase from 4.31 to 4.38 g/cm^3. Given the theoretical density of titanium (4.51 g/cm^3), the possibility of cavity formation within the spherical particles were investigated by qualitative cross-section secondary electron Scanning Electron Microscopy (SEM) analysis. Quantitative determination was not possible due to sample availability.

Due to the smooth and spherical nature of the sample powder, individual spheroidised particles were observed to regularly pull out of typical epoxy resins used during the first grinding step of metallographic sample preparation. Therefore, cross-sections of the sample powders before and after plasma treatment were mounted in strongly adhesive Pratley Steel Quickset®. SEM as well as EDS (Energy Dispersive Spectroscopy) analysis were conducted at Necsa using a Quanta 200 3D (at 30 kV and 2.6 nA) with an SD detector.

Particles were either internally fully dense (Fig. 1), contained voids (Fig. 2) or were completely hollow (Fig. 3). The mechanism by which these flaws were introduced was after complete melting and rapid quenching when exiting the plasma. The shell of some particles solidified whilst the centre remained in liquid phase and at a lower density, forming a cavity during complete cooling of the particle. Treating large particles might result in the formation of enlarged cavities[3]. Therefore, the plasma operation parameters need to be adjusted to increase the particle residence time to allow for controlled cooling during solidification of the melted particles to reduce the temperature difference between the core and the outer shell of the particle.

References:
1. Williams, J.V. and Revington, P.J. (2010) Int. Assoc. of Oral and Maxillofacial Surgeons 39, 182.
2. Boulos, M.I. (1985) Pure and Appl. Chem. 57, 1321.
3. Gignard, N.M. (1998) PhD Thesis, National Library of Canada.

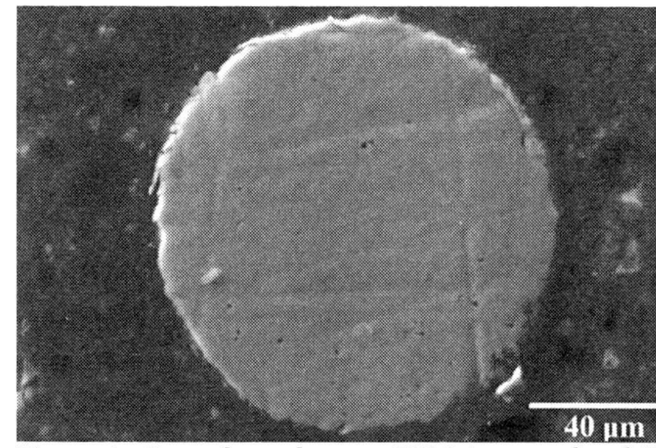

Figure 1. Secondary electron SEM image of a fully dense Ti metal particle after plasma treatment.

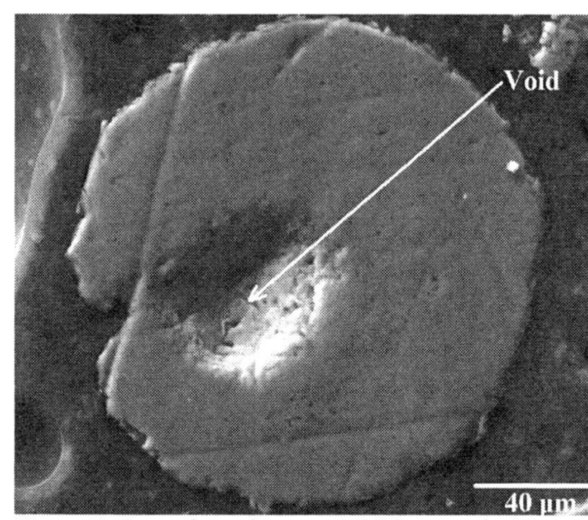

Figure 2. Secondary electron SEM image of a Ti metal particle containing a void after plasma treatment.

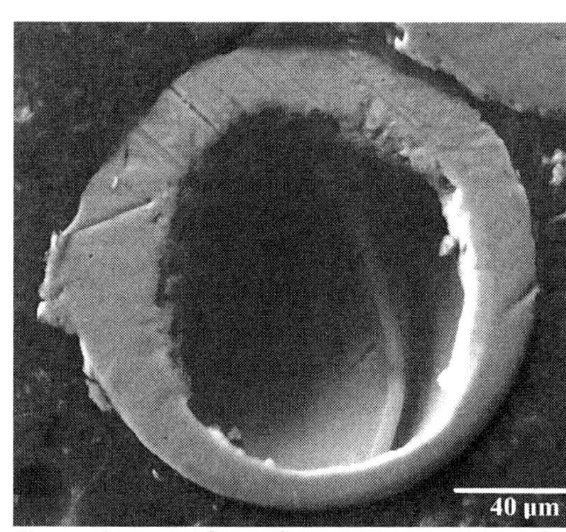

Figure 3. Secondary electron SEM image of a completely hollow Ti metal particle found after plasma treatment.

Corresponding author: Ryno.vandermerwe@necsa.co.za

MICROSCOPIC AND FLUORESCENCE PROPERTIES OF CARBON DOTS SUPPORTED ONTO SILICA AND HOLLOW CARBON SPHERES

O. Mkhari[1], E. Nxumalo[2], N.J. Coville[1,3] and M.S. Maubane-Nkadimeng[1,3,4]

[1]Molecular Sciences Institute, School of Chemistry, University of the Witwatersrand, Johannesburg, [2]Nanotechnology and water unit, University of South Africa, Pretoria, [3]DST-NRF Centre of Excellence in Strong Materials, University of the Witwatersrand, Johannesburg, [4]Microscopy and Microanalysis Unit, University of the Witwatersrand,Johannesburg

The discovery of carbon dots (Cdots) as a new type of carbon based nanomaterial with size below 10 nm has attracted many scientific inverstigations.[1] This relates to their inexpensive preparation methods to give materials with a range of surface substituents, low toxicity, good electron transfer rates and biocompatibility, which make it attractive for use in many applications.[2] The most interesting property of Cdots is photoluminescence which depends on the starting materials and can also be tuned through doping.[3] However, there are several key issues related to Cdots that still need to be addressed i.e. purification steps and poor yields. As such, supporting Cdots on various surfaces can be used to minimise the above issues.

Supported Cdots have been reported to have improved properties such as light capturing ability and improved electron transfer rates relative to their unsupported counterparts. Supported Cdots are promising materials as reagents for use in chemical processes such as adsorption and drug delivery.[4,5] In this study, we have evaluated the fluorescence and microscopic properties of Cdots supported onto the surface of silica spheres (SSs) and hollow carbon spheres (HCSs).

Synthesis of Cdots was achieved using a hydrothermal method at 200 °C for 3 h in an autoclave reactor. The SSs were synthesized using a modified Stober method, while HSCs were obtained by using the as-synthesized SSs as template.[6] The Cdots were supported onto the surface of SSs and HCSs using reflux (80 °C) and sonication (at room temperature), conditions respectively.

The photoluminescence properties of the Cdots and Cdots supported on SSs and HCSs were investigated (Figure 2). The maximum emission of the Cdots, Cdots/SSs and Cdots/HCSs were observed at 510 nm, 440 nm and 418 nm when excitation wavelength of 440 nm, 380 nm and 340 nm were used, respectively. The excitation wavelength dependent nature of the Cdots was observed by photoluminescence spectroscopy in the 350 to 500 nm excitation wavelength range. The Cdots framework was evaluated using RAMAN spectroscopy; the G-band and D-band were observed at 1566 cm^{-1} and 1358 cm^{-1} respectively.

In conclusion, Cdots were successfully synthesized by a hydrothermal method in an autoclave reactor using chitosan as carbon and nitrogen source. The photoluminescence properties of Cdots were tuned by using different supports, as seen by the change in the maximum emission of bare Cdots and Cdots supported onto SSs and HCSs. The changes in the maximum emissions can open doors for various applications of Cdots supported on SSs and HCSs.

References:
1. Yubin, S. *et al.* (2015) J. Mater. Chem. C 3, 5976.
2. Reckmeier, C. J. *et al.* (2017) Chem. Mater 29, 10352
3. Zhou, D. *et al.* (2017) Chem. Mater 29, 1779
4. Li, Y. *el al.* (2018) Talanta. 180, 18
5. Zhang, B. *et al.* (2015) *Colloids Surfaces A Physicochem. Eng. Asp.* 485, 34
6. Stober, W. *et al.* *(1968)* J. Colloid Interface Sci. 26, 62

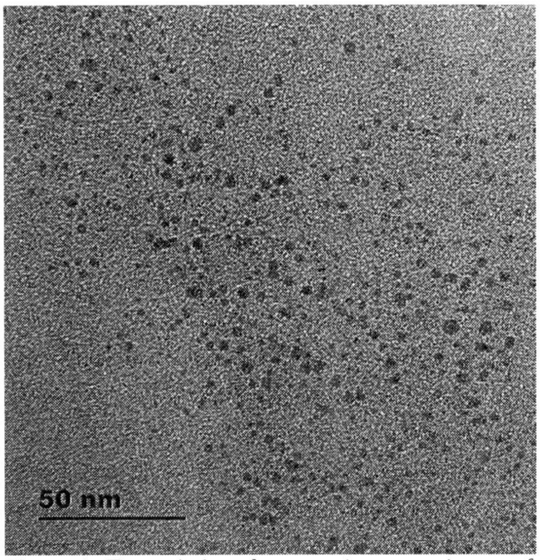

Figure 1. Transmission electron microscopy image of Cdots

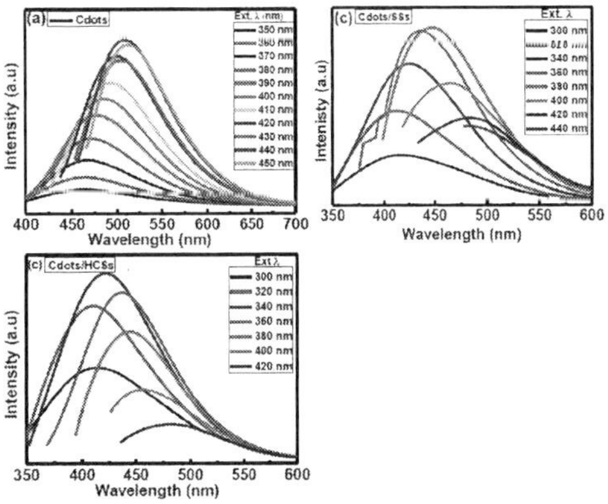

Figure 2. Optical properties of Cdots, Cdots/SSs and Cdots/HCSs

Corresponding author: 1901046@students.wits.ac.za

MICROSCOPY INVESTIGATION OF SINTER-INFILTRATION OF Ti6Al4V BLENDS

P. Govender and D.C. Blaine

Department of Mechanical & Mechatronic Engineering, University of Stellenbosch, Stellenbosch

Sintering is a high temperature process through which powders bond together to form a continuous solid material at temperatures below their melting points[1]. Infiltration is the process of filling the pores of a sintered part with a low melting point metal to improve tensile strength, hardness, etc[2]. This technique is commonly used for ferrous powder metallurgy alloys by infiltrating them with molten copper during sintering. Sintering theory indicates that powder materials that are sintered to a fractional density above 92% no longer have open pore channels and thus would not be suitable for infiltration. In this study, a similar process for infiltrating sintered Ti6Al4V with aluminium is designed and evaluated.

Two Ti6AlV powder blends were mixed and die compacted into ø10x10 mm right cylinders at 400 MPa. These powder blends consist of a commercially pure Ti powder mixed with a 60Al-40V master alloy powder (CPTi+MA), and an elemental blend of Ti, Al (6 wt%) and V (4wt%) powders (BE). The powder compacts were sintered to peak temperatures of 1100 ºC and 1200 ºC, respectively, with a dwell time of 2 hours at the peak temperature. They were sintered under vacuum in a horizontal tube furnace. Thereafter, pure Al discs were placed on top of the sintered Ti6Al4V sample, while heating the compact and sample under N_2 to 700-900 °C for 30min-1 hour dwell times. The purpose of this process step is to melt the Al. Initially, Al powder was compacted into discs for this purpose, but when these discs did not melt, slices from a wrought Al rod were used in order to determine whether melting was prohibited by the typical passive oxide layer[3]. Additionally, the infiltration heat treatment takes place in a N_2 atmosphere due to a low dew point, as moisture is detrimental in heat treatments of Al and it has been reported that this allows for the formation of AlN on the particle surface which disrupts the persistent oxide layer and thus aids in infiltration[3]. Samples were sectioned, mounted and polished after sintering and also after the infiltration step. An Energy Dispersive X-ray spectrometer (EDS) was used for imaging and elemental mapping.

The BE samples showed negligible densification during sintering remaining at 75%. In contrast, the CPTi+MA samples densified from a green relative density of 75% to 85-90%. After the infiltration heat treatment it was found that the Al compacts did not fully melt for either powder blend, thus no infiltration took place.

EDS analysis of the sintered Ti6Al4V samples was performed. Fig. 1 shows the elemental maps of the CPTi+MA samples after sintering. A homogenous distribution of elements is observed, yet pockets of high V concentration are apparent. Al has a high solubility in Ti, and therefore it is probable that Al leached out of the MA powder particles into the Ti powder particles. Fig. 2 shows the elemental maps of the BE samples after sintering. Large pores are revealed, with high Al and N concentrations close to the pore surface. The high Al concentration indicates that individual Al powder particles melted during sintering and quickly diffused into the Ti and V powders, leaving large pores behind. Large cracks were found in all BE samples, which appear to have formed during metallographic preparation. The coincidence of high concentrations of Al and N at the pore surfaces points to the formation of AlN, as reported in studies where Al is sintered in a N_2 atmosphere[3].

EDS analysis of the Al disk after the infiltration heat treatment showed that Ti and V diffused out of the Ti6Al4V into the Al disks. Diffusion of these elements into the Al may increase the alloyed Al's melting temperature giving reason as to why the Al did not melt.

CPTi+MA and BE powder blends of Ti6AlV were mixed, compacted, sintered and infiltrated. Infiltration did not occur largely due to persistent oxide films on Al, as well as Ti and V diffusing into the Al disks. It was seen the compacts started to melt upon contact yet did not fully melt throughout. Melting and subsequent liquid phase diffusion of Al during sintering leads to formation of pores in the BE samples and inhomogeneous V distributions.

References:
1. Schaschke, C., (2014) Dictionary of Chemical Engineering, Oxford University Press, 346.
2. Bosman, H.L. and Blaine, D.C. (2016) Influence of powder particle size distribution on press-and-sinter titanium and Ti-6Al-4V preforms, (Masters thesis) SUNScholar http://hdl.handle.net/10019.1/98709, Stellenbosch, 44.
3. Schaffer, G.B. *et al.* (2005) Acta Mater. 54(1), 131.

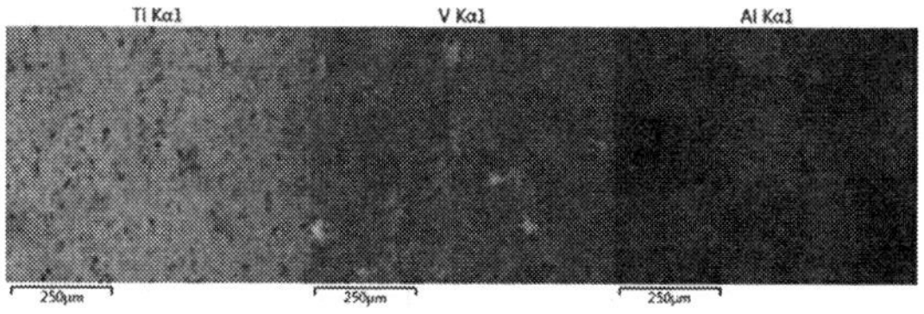

Figure 1. EDS mapping of CPTi +MA sample after infiltration attempt

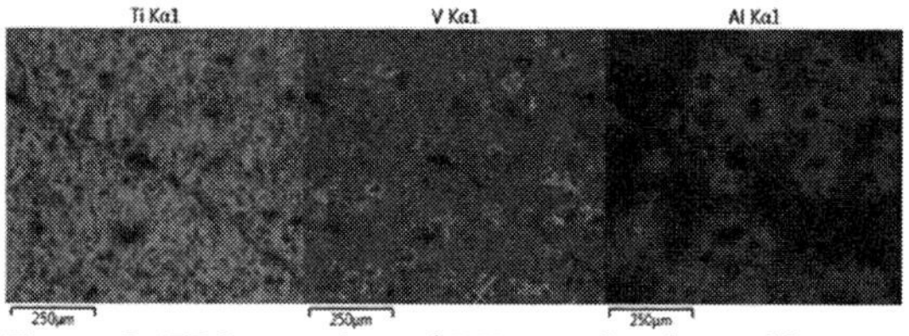

Figure 2. EDS mapping of BE sample after infiltration attempt

Corresponding author: 18375391@sun.ac.za

MICROSTRUCTURAL ANALYSIS OF Ti-Cu ALLOYS FOR DENTAL APPLICATIONS

N.D.E. Masia[1,2,3], L.H. Chown[2,3], M. Smit[1], A. Mwamba[1,3] and L.A. Cornish[2,3]

[1]Advanced Materials Division, Mintek, Randburg, [2]School of Chemical and Metallurgical Engineering, UW, JHB, [3]DST-NRF Centre of Excellence in Strong Materials, hosted by the UW, JHB

Titanium has been the material of choice in several disciplines of dentistry due to its excellent characteristics such as biocompatibility, osseo-integration, high wear and corrosion resistance, low compatibility issues and high strength. Corrosion of titanium dental implants causes implant failure and can trigger peri-implantitis[1], which is caused by bacteria. Bacterial adhesion on the implant surface can be avoided by hydrophilic coatings, but these can inhibit tissue integration[2]. Antibiotics can also be used against peri-implantitis but long-term high doses may result in deleterious side-effects and bacterial resistance[3]. Adding copper to titanium alloys increases antibacterial properties and retains adequate biocompatibility[4]. The shape and size of Ti_2Cu phase affects antibacterial properties and mechanical properties, as well as corrosion resistance[5]. Annealing and ageing could significantly improve the mechanical properties, corrosion resistance and antibacterial rate due to the redistribution of copper and size distribution of Ti_2Cu precipitates. This study assessed the effect of Ti_2Cu and its proportions on the corrosion resistance, compared to commercial pure titanium (CP-Ti, grade 4).

The Thermo-Calc program with TTTI3 (Ti-alloy database) was used to predict the phases. Ti-Cu alloys of different compositions, 15, 25, and 47 wt % Cu, were produced in a button arc melting furnace. They were studied in the as-cast and annealed conditions (900°C water quenched). The microstructures were observed using visible light microscopy. Compositions were determined using an electron probe microanalyser, and the phases were identified by X-ray diffraction. Corrosion resistance was done by potentiodynamic polarisation in a phosphate buffered saline solution at 37°C at 7.4 pH while purging with nitrogen gas.

Thermo-Calc predicted only equilibrium alpha, beta and Ti_2Cu while XRD found Ti_3Cu in Ti-25Cu in the as-cast condition and not after annealing. Figs. 1-2 shows the microstructures. CP-Ti consisted of a lamellar αTi microstructure. Ti-15Cu and Ti-25Cu had similar microstructures consisting of eutectoid (lamellae) αTi and Ti_2Cu with αTi islands; except that XRD indicated the presence of Ti_3Cu instead of Ti_2Cu in as-cast Ti-25Cu. Ti-47Cu consisted of a dual phase Ti_2Cu and TiCu structure with some copper. In the cast alloys the addition of copper decreased the corrosion resistance from 0.00059 to 0.00135 mm/y; while the annealed alloys showed an improvement in the corrosion resistance from 0.00099 to 0.000029 mm/y. The corrosion rates were well within the acceptable range (0.02-0.13 mm/y) for biomaterial design and application.

References:
1. Rodrigues, D. *et al.* (2013) Mat. 6(11) 5258.
2. Francolini, I. and Donelli, G. (2010) FEMS Imm. & Med. Microbiol. 59(3), 227.
3. Andrews, J. (2002) J. Antimicrob. Chemo. 49(6),1049.
4. Osório, W. *et al.* (2010) Electrochim. Acta 55(10), 759.
5. Zhang, E. *et al.* (2016) Mat. Sci. & Eng. 69, 1210.

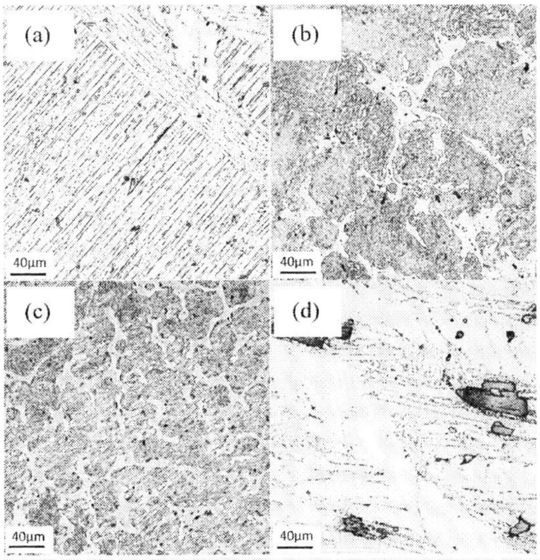

Figure 1. As-cast Ti-Cu alloy microstructures: (a) CP-Ti (b) Ti-15Cu (c) Ti-25Cu (d) Ti-47Cu.

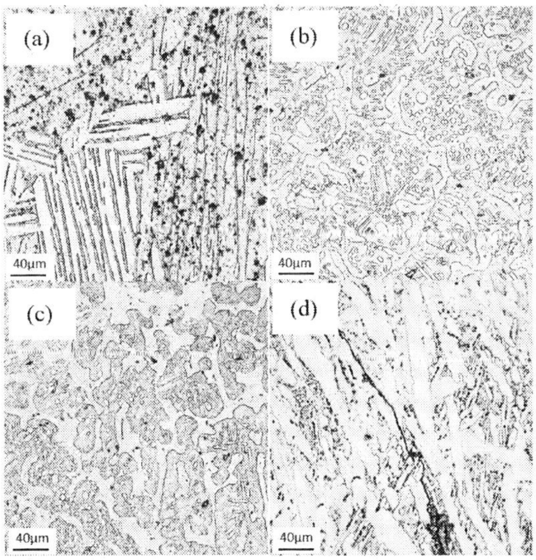

Figure 2. Annealed Ti-Cu alloy microstructures: (a) CP-Ti (b) Ti-15Cu (c) Ti-25Cu (d) Ti-47Cu.

Corresponding author: nomsombulukoh@mintek.co.za

MICROSTRUCTURAL AND CHEMICAL CHARACTERISATION OF SPARK PLASMA SINTERED Ti-Ta-Zr POWDERS

B.L. Bayode[1], M.L. Teffo[2], R. Machaka[3] and P.A. Olubambi[1]

[1]Centre for Nanoengineering and Tribocorrosion, School of Mining, Metallurgical and Chemical Engineering, UJ, Johannesburg, [2]Institute of NanoEngineering Research, Department of Chemical, Metallurgical and Materials Engineering, TUT, Pretoria, [3]CSIR, Pretoria

Suitable Ti-alloys containing dissolved Ta and Zr elements are used for applications in biomedical and aerospace technologies due to their attractive properties including high biocompatibility, high strength-to-weight ratio, and shape memory behaviour. Their shape memory property is due to the presence of martensitic structures which undergo reversible martensitic transformation[1].

The control of grain size and size distribution, and compositional homogeneity remains a challenge in the fabrication of Ti-Ta-Zr alloys by conventional fabrication techniques such as casting[2]. This limitation can be minimised by spark plasma sintering (SPS), a novel powder metallurgy technique, which involves the use of predetermined powder formulations with desirable compositions and blends of particle sizes[1]. However, SPS is also plagued with the challenge of achieving full densification and effective metallurgical bonding of particles[2]. This study presents the microstructural, chemical and phase characteristics of spark plasma sintered (SPSed) Ti-(30-x)Ta-xZr alloys (x = 0, 5, 10 at.%) with particle sizes ranging from 2-29 µm.

The SPS parameters include sintering temperature (1200 °C), heating rate (100 °C/min), pressure (30 MPa), and sintering dwell time (10 mins). As-sintered specimens were polished using standard grinding and polishing plates and cloths with diamond suspension pastes down to 1 µm sizes. A field emission scanning electron microscope (JEOL JSM-7600F), equipped with energy dispersive X-ray spectroscopy (Oxford Instrument X-Max with INCA software), operated at 25 kV accelerating voltage, a probe current of 14 µA and a working distance of 6 mm was used for the microstructural analysis. Phase identification of sintered specimens was done on a PANalytical Empyrian multi-platform X-ray diffractometer.

In Fig. 1a-c, the regions labelled I contained α-Ti phase with twinned martensite structures. EDX analysis of these regions revealed that they contained the highest concentration of Ti elements compared to other regions, with a concentration of Ta less than 9 at.%. The regions labelled II were also Ti-rich regions of ß-Ti but with a slightly higher concentration of Ta (above 9 at.%). Studies have established that at above 10 at.%, Ta solutes will stabilise ß-Ti phase at low temperatures. The third region labelled III is the ß-Ta phase, which contained the highest concentration of Ta (above 50 at. %). The microstructural characterisation showed effective metallurgical bonding between the particles, little or no surface porosity (implying high densification) and significant inter-diffusion of atoms. Although a homogeneous chemistry was not recorded, the grain sizes were the same as the initial particle size distribution of starting powders. These results indicate the effectiveness of the SPS technique for the fabrication of Ti-Ta-Zr alloys for useful aerospace applications.

Financial support from the National Research Foundation (NRF) of South Africa and the National Aerospace Centre (NAC.1314.001) is acknowledged.

References:
1. Hussein, M.A., Suryanarayana, C. and Al-Aqeeli, N. (2015) Materials and Design 87, 693.
2. Ismail, M.H. *et al.* (2012) Materials Letters 70 , 142.

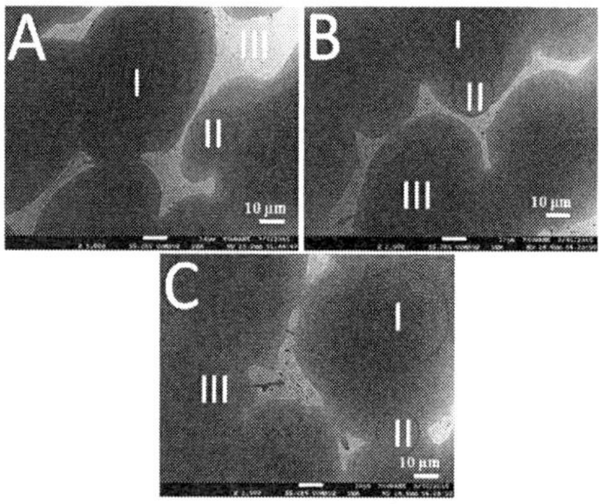

Figure 1. Z-contrast images of sintered products obtained from SEM BSE: (A) Ti-30Ta (40 MPa), (B) Ti-25Ta-5Zr and (C) Ti-20Ta-10Zr.

Corresponding author: 217072547@student.uj.ac.za

MICROSTRUCTURAL CHARACTERISATION OF SWIFT HEAVY ION INDUCED CRYSTAL ROTATION IN SINGLE CRYSTAL NiO

D.A. Venter and J.H. O'Connell

Centre for HRTEM, Nelson Mandela University, Port Elizabeth

Swift heavy ions (SHI) are characterised by having energies >1 MeV per nucleon with an atomic mass Z > 6. Exposure to irradiation of this type has been known to produce latent tracks in insulators. The energy loss through the material (dE/dx) is mostly due to electronic excitation processes[1]. The extent of the track formation and microstructural changes induced has been seen to be material specific. Models such as the thermal spike model have been used to explain these changes, however rough approximations are still made[2-3]. A particularly under studied phenomena is the collective rotation seen in materials exposed to off-normal incidence irradiation. To date the only available experimental data is X-ray diffraction studies which does not allow for depth-dependent rotation studies and very little can be deduced regarding the microstructure of the modified volume.

In this study, the microstructural changes in a single crystal NiO(100), irradiated at 45° with 593 MeV Au to a fluence of 1.4×10^{14} ions/cm^2 has been studied using electron microscopy techniques. A depth dependent investigation of the crystal rotation was performed using electron backscatter diffraction (EBSD), transmission Kikuchi diffraction (TKD) and selected area electron diffraction (SAED). High-resolution transmission electron microscopy (HRTEM) was utilised to interrogate the microstructural changes facilitating bulk rotation.

Maximum relative rotation was measured as 24.7 ± 2.0° (EBSD), 24.3 ± 2.0° (TKD) and 23.0 ± 0.2° (SAED). High-resolution transmission electron microscopy (HRTEM) was utilised to interrogate the microstructural changes. Approximately 2% of the incoming ions produced streaks of lower density material that were seen in the damaged regions. Rotation extends to 10 µm below the irradiated surface suggesting a threshold electronic stopping power of around 10.5 keV/nm for crystal rotation.

The slip system for NiO is the same as that for the NaCl system[4]. {110} planes slip preferably along the <110> direction[5]. Slip is known to cause a gradual rotation in the crystal lattice. This is facilitated by the crystal extending and the orientation of the crystal changing relative to the tensile axis[6]. The crystals tensile stress was found to exceed the critical resolved shear stress in the regions of 6.8 µm to 10.2 µm below the surface, where the rotation was observed. This suggests that slip occurred to facilitate the crystal rotation.

Financial assistance of the National Research Foundation (NRF) of South Africa, (grant no. 112517) as well as that of Nelson Mandela University is gratefully acknowledged. A special thank you to Siegfried Klaumünzer, for providing the samples used in this study.

References:

1. Meftah, A. *et al.* (1993) Phys. Rev. B, 48(2), 920.
2. Dufour, C. and Toulemonde, M. (2016) Ion Beam Modification of Solids. Switzerland, Springer International Publishing.
3. Toulemonde, M., Bouffard, S. and Studer, F. (1994) Nucl. Inst. Methods Phys. Res. B, 91(1–4), 108.
4. Wang, J., Fisher, E. and Manghnzmi, M. (1991) Chinese Phys. Lett. 8(3), 153.
5. Evans, A.G., Rajdev, D. and Douglass, D.L. (1972) Oxid. Met. 4(3), 151.
6. Hosford, W.F. (2010) Mechanical Behavior of Materials, America, Cambridge University Press.

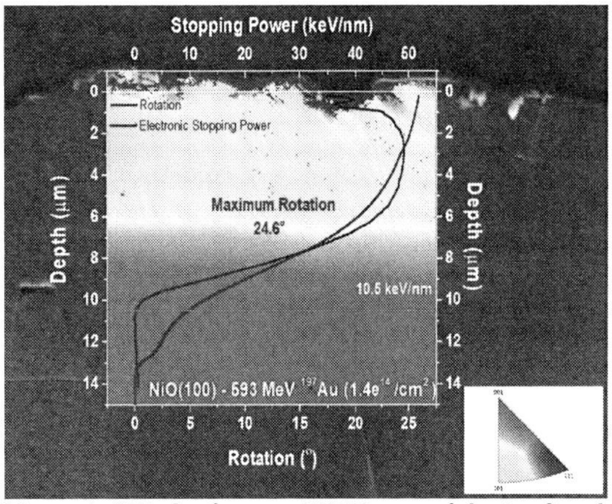

Figure 1. IPF plot from an EBSD map of the irradiated specimen seen in cross section. Ion irradiation direction was at 45° from the top right of the image. The overlay shows a plot of the relative rotation of the structure and the calculated electronic stopping power as a function of the depth.

Corresponding author: s212244450@mandela.ac.za

MICROSTRUCTURAL DETORIATION ASSESSMENT OF EX-SERVICE POWER PLANT STEEL USING CONCENTRIC BACKSCATTER IMAGING

M. van Rooyen[1], T.H. Becker[1], J.E. Westraadt[2] and G. Marx[2]

[1]Materials Engineering, University of Stellenbosch, Stellenbosch, [2]Centre for HRTEM, NMU, Port Elizabeth

Successful operation of coal-fired power plants requires monitoring of the material health of constituent critical components that are exposed to high temperatures and stresses over extended operating lives (> 25 years)[1]. The microstructure of these steels, such as X20CrMoV12-1 tempered martensite ferritic alloy, undergo deterioration during service due to creep that degrade their mechanical properties[2]. Plant repair and replacement strategies are heavily dependent on the microstructural damage assessment of the exposed alloys. This study seeks to demonstrate the use of Concentric Backscatter (CBS) imaging as a comprehensive diagnostic tool for microstructural damage assessment of two different service-exposed X20 steels. Following CBS analysis, the resulting accelerated creep behavior is assessed using Digital Image Correlation (DIC) strain measurement.

Two ex-service X20 piping steels were supplied from a local power utility for this study, with damage classes identified from surface replica cavity counts using visible light microscopy. The high damage X20 (200-600 cavities/mm) served for 17 years at a pressure of 18.1 MPa and temperature of 543 °C. Low damage X20 (60-90 cavities/mm) operated at 15 years at 17 MPa and 545 °C. Thin foils were prepared from colloidal silica polished bulk specimen 3 mm discs using a Struers TenuPol-5-twin-jet electropolisher (5 % $HClO_4$ solution; 21 – 30 V; -20°C). CBS images were acquired across a collective area of 600 μm^2 using a detector attached to a FEI Helios NanoLab DualBeam 650 focused ion beam scanning electron microscope (SEM) at 5 kV accelerating voltage, 0.20 nA probe current and 4 mm working distance. Micro-grains, $Cr_{23}C_6$ and Laves (Mo-rich) phase precipitates (identified by the red outline, light grey and white speckles in Fig. 1, respectively) morphological details were obtained from a thresholding routine applied to the micrographs using MIPAR v. 2.0 software[3].

An unpaired t-test statistical analysis was performed on the morphological measurements of the microstructure and main material differences are summarised in Table 1. Clearly, smaller, more elongated $Cr_{23}C_6$ particles, lower number densities of Laves phase and shorter widths in micro-grains were found in the high damage X20 as opposed to the low state. This indicates creep-driven microstructural degradation which is manifested as higher creep strain rates as measured by LaVision Stereo-DIC (61 pixel^2 subset size and 15 pixel step size) during accelerated creep testing on a Gleeble thermomechanical simulator as shown for various temperatures in Fig. 2[2]. These findings therefore indicate that clear indicators of creep damage in service-exposed X20 plant steel are identified through observation of key microstructural features within a single microscopy technique, namely CBS imaging.

References:
1. Aghajani, A., Somsen, C.H. and Eggeler, G. (2009) Acta Materialia 57(17), 1.
2. van Rooyen, M. *et al.* (2019) Materials 12(19), 3106.
3. Marx, G. (2019) Microstructural evolution of welded creep aged 12 % Cr martensitic stainless steel (PhD thesis), NMU.

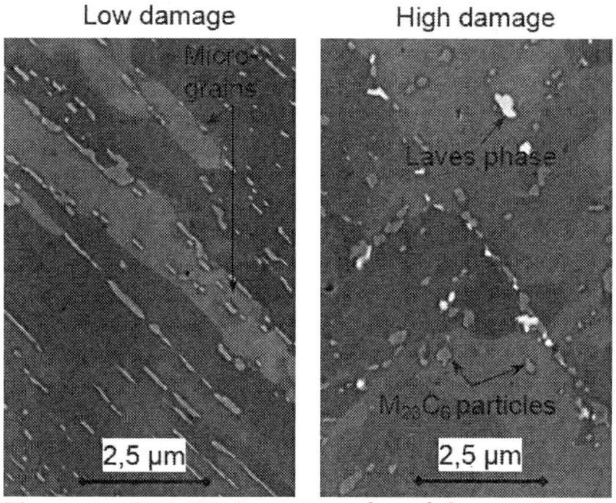

Figure 1. CBS-SEM micrographs of low and high damage X20 showing precipitates and micro-grains.

Table 1. Statistically different microstructural parameters.

Damage state	$Cr_{23}C_6$ size (nm)	Laves volume density (µm^{-3})	Micro-grains size (µm)
Low	182 ± 12	0.4 ± 0.4	0.54 ± 0,02
High	196 ± 2	1,9 ± 0,7	0.75 ± 0,04

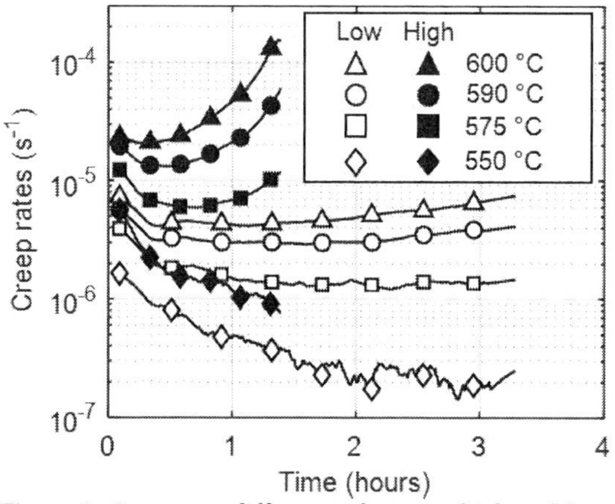

Figure 2. Creep rate differences between high and low damage X20 as measured by DIC at several temperatures.

Corresponding author: melzvanrooyen@sun.ac.za

MICROSTRUCTURAL EVOLUTION OF TiN REINFORCED Ti-6Al-4V NANOCOMPOSITE

M.E. Maja, O.E. Falodun, B.A. Obadele and P.A. Olubambi

Centre for NanoEngineering and Tribocorrosion, School of Mining, Metallurgy and Chemical Engineering, UJ, JHB

Spark plasma sintering (SPS) is a process known for sintering at lower temperatures in shorter processing times that result in improved properties compared to the traditional sintering processes[1]. This process makes it possible to successfully sinter complex nanocomposites such as TiN reinforced Ti–6Al–4V, which would be difficult to fabricate using conventional hot pressing methods. Ti-6Al-4V is of interest due to its excellent mechanical properties at temperatures below 400°C, while TiN is used for its good thermal stability and strength at elevated temperatures. This work focuses on analysing the morphology of spark plasma sintered TiN reinforced Ti–6Al–4V, to widen the application of Ti-6Al-4V at elevated temperatures.

Ti–6Al–4V powder (APS 25 µm and purity of 99.9%) was reinforced with volume fractions of 1-4 vol. % TiN powder (20 nm and purity of 97%) using the spark plasma sintering technique, while the phase and crystallographic orientations were characterised using a high–resolution light microscope and a field emission scanning electron microscope (FE-SEM) equipped with EDS/EBSD. The grain sizes were obtained from the EBSD results.

Microstructural analysis of Ti-6Al-4V, shown in Fig. 1(a), revealed a fully lamellar microstructure comprising two distinct α and β phases (on the Ti-6Al-4V without any reinforcement). Fig. 1(b), Ti-6Al-4V reinforced with 1-4 vol. % TiN, showed a bimodal structure comprising evenly distributed interconnected primary α grains and a transformed β phase. TiN nanoparticles were observed along the grain boundary and α phase.

Twinning was observed, and may have been caused by insufficient slip planes and by the fibre strengthening which refined the grain size[2]. EBSD maps showed an increase of the α phase from 78% to 92%, while the β phase decreased to 1.9% as a result of increased vol% TiN. The interstitial element has a strong stabilising effect on the α phase, which ultimately raises the transus temperature[3].

The average grain size shown in Fig. 2 decreased with an increase in the amount (vol %) of TiN, except for a slight increase in Ti-6Al-4V + 4 Vol% TiN. The significant decrease in grain size could be due to the presence of α stabiliser at the grain boundaries that acts as a solid solution strengthener. The TiN in Ti-6Al-4V restricts grain growth and promotes grain refinement[4]. The average grain size of Ti-6Al-4V + 4 vol% TiN increased slight from 2.55 to 3.65 µm which could be due to saturation of TiN on the grain boundary and colonies of TiN at the grain boundary.

Reinforcing the Ti-6Al-4V matrix with TiN nano-composites transforms and refines the grain structure from fully lamellar to bimodal which could improve mechanical properties.

References:

1. Torralba, J.M. (2014) Comprehensive Materials Processing. Amsterdam, Elsevier.
2. Guo, Y.F. *et al.* (2014) Size effect on deformation twinning at the nanoscale in hcp metals. Beijing Jiaotong University, Institute of Engineering Mechanics.
3. Pederson, R. (2002) Microstructure and Phase transformation of Ti-6Al-4V. Sweden, Luleå tekniska universitet.
4. Moorhouse, B. (2013) Controlling the interstitial element concentration in Ti-6Al-4V using calciothermic reduction. London, Imperial College.
5. Kao, Y.L. *et al.* (2005) Mater. Sci. Eng. A 398(1-2), 93.
6. Lutjering, G. and Williams, J.C. (2003) Titanium, Engineering Material Processes. New York, Springer.

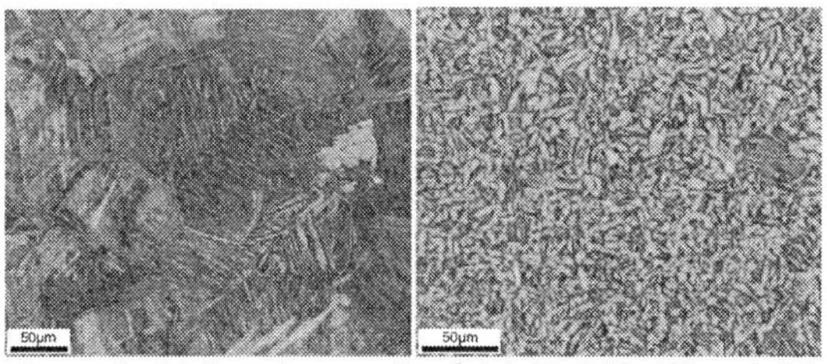

Figure 1. Light micrograph of (a) Ti–6Al–4V and (b) 1% TiN reinforced Ti–6Al–4V

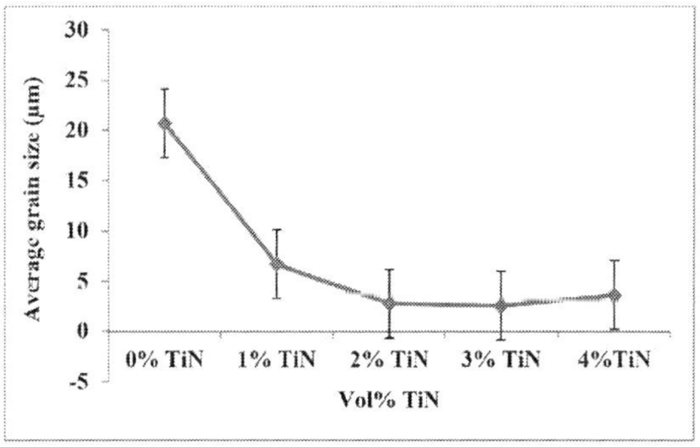

Figure 2. Average grain size as a function of vol% TiN

Corresponding author: mosimaedith@gmail.com

MICROSTRUCTURAL INVESTIGATION IN SEVERELY DEFORMED AA5182 ALUMINIUM ALLOY

G. Morrison[1,2], S. George[1] and C. Lang[2]

[1]Centre for Materials Engineering, University of Cape Town, [2]School of Engineering, Macquarie University, NSW, Australia

Aluminium alloys are used in the beverage can industry due to their high strength to weight ratios, and good recyclability[1]. The AA5182 aluminium alloy is used for beverage can ends. AA5182 is an alloy of aluminium with the main alloying elements of magnesium, iron, manganese, and silica.

AA5182 beverage can end sheet is produced by ingot casting, hot rolling, cold rolling, and coil coating. This can end stock (CES) is rolled down to a thin sheet of approximately 0.2 mm thick. During the sheet production processes, the CES is exposed to heat treatments, and severe deformation during the rolling processes. The heat treatments and strain cause changes in the crystallographic texture, and the microstructure of the CES[2]. These changes primarily result from the arrangement of dislocations in the subgrains of the CES, and the orientation of grains, and occur in a non-uniform fashion in the material[3]. The aim of this research is to determine if the microstructure can be measured in severely deformed AA5182.

AA5182 CES was received from Hulamin Rolled Products in a number of severely cold worked conditions, after coil coating. In order to avoid interference in the scanning electron microscope (SEM), the non-conductive coil coating was removed using sulfuric acid. CES samples of approximately 10 mm x 10 mm were cut from the sheets, and mounted on SEM stubs using silver paste.

The SEM stub holders were placed in a FEI Helios Nanolab 650 SEM at the Centre for HRTEM at the Nelson Mandela University. The SEM had a gallium ion, focused ion beam (FIB), attachment. The CES samples were cut from the desired transverse direction (TD) plane, using the FIB technique. The FIB cuts were each milled from the desired regions of the CES. The FIB cuts were then attached to different fingers on the copper holders, using carbon deposition. These FIB cuts were then thinned down to below 100 nm, to be able to be later examined using transmission electron microscopy (TEM). The transmission Kikuchi diffraction (TKD) technique was used to resolve the finer features of the CES microstructure and texture. Once the optimum settings for the AA5182 alloy were determined, the FIB cuts were investigated. The JEOL JSM-7001F SEM was used to perform the TKD maps, using a stepsize of 0.02 µm and a 20 kV accelerating voltage.

SEM images and TKD maps were obtained for the CES, as shown in the example TKD map in Fig. 1. The average grain sizes and misorientation angles, for all three CES conditions, were determined using the HKL Tango post-processing software. The pole figures were also measured for all three CES conditions, using the HKL Mambo post-processing software. The results from the post-processing of the TKD maps and SEM images, showed that the size and shape of the grains in the samples could be measured. When the TKD data was analyzed with HKL Mambo, the difference between the three conditions was more pronounced. Although the data showed differences in the preferred crystallographic orientations of the different CES conditions, the TKD analysis area was not representative of the bulk samples, as it used too few grains. The FIB cut samples which had not been exposed to TKD, will be used to determine the microstructure in more detail using TEM. X-ray diffraction (XRD) will be used to determine the bulk texture in larger samples than the FIB cuts.

The TKD technique was useful to demonstrate the CES microstructures and crystallographic texture, in the form of different grain sizes, and preferred orientations. While the TKD results are useful, future TKD work is needed on aluminium CES in a number of other conditions. An investigation of the types of subgrain structures, using the TEM technique, and the bulk crystallographic texture, using X-ray Diffraction (XRD), on the CES samples, will also give a more complete picture of the AA5182 CES.

References:
1. Yu, X. *et al.* (2005). Mater. Sci. Eng. A 394(1), 376.
2. Crumbach, M. *et al.* (2006) Model. Simul. Mater. Sci. Eng. 14(5), 835.
3. Banabic, D. and Siegert, K. (2004) CIRP Ann. 53(1), 219.

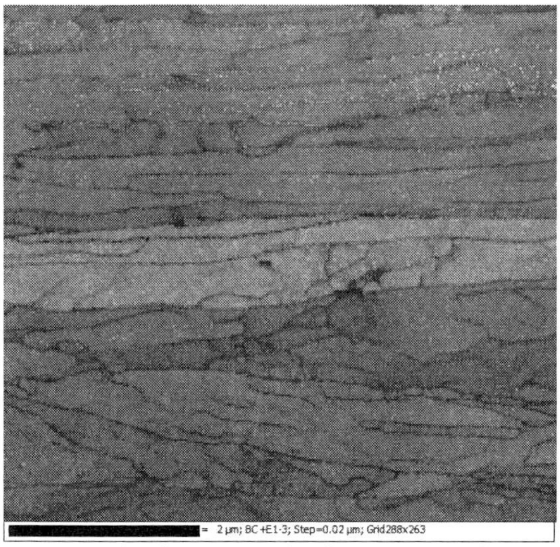

Figure 1. A band contrast and Euler TKD map of a CES sample.

Corresponding author: mrrgra008@myuct.ac.za

MICROSTRUCTURE EVOLUTION IN HYDROGEN TEMPERED AND NON-HYDROGEN TEMPERED Ti-6Al-4V MARTENSITE

V.N. Vilane[1], R.D. Knutsen[1] and J.E. Westraadt[2]

[1]Centre for Materials Engineering, Department of Mechanical Engineering, University of Cape Town, Cape Town,
[2]Centre for HRTEM, Nelson Mandela University, Port Elizabeth

Hydrogen treatment can refine cast Ti-6Al-4V microstructures and improve mechanical performance. However, the performance of hydrogen treated microstructures is often not evaluated. Consequently, it is not widely known that hydrogen treated microstructures do not always yield improved performance. In contrast, hydrogen tempering contributes to ductility loss in tempered martensite[1] because it promotes titanium aluminide (Ti_3Al) formation. The contribution of hydrogen to the promotion of Ti_3Al precipitation in tempered martensite has not yet been articulated. This study articulates the contribution of hydrogen to tempered martensite evolution and Ti_3Al formation.

Wrought Ti-6Al-4V was converted to simulated cast (Sim. Cast) Ti-6Al-4V by annealing at 1100 °C for 2 hours. Sim. Cast samples were hydrogenated to 20 at. %H at 650 °C, solution treated at 900 °C and quenched, aged (or hydrogen tempered) at 580 °C and dehydrogenated at 675 °C or 750 °C. Non-hydrogen tempered (control) samples were similarly prepared in a vacuum atmosphere but solution treated at 1080 °C. Electron backscatter diffraction (EBSD) and Energy Dispersive Spectroscopy (EDS) were performed in the SEM (JEOL 7100). Thin TEM foils were extracted from sites of interest using a focused ion beam in the Helios Nanolab 650. Subsequent high angle annular dark field (HAADF) imaging was performed in scanning-TEM mode (STEM) in a Cs corrected ARM 200F.

In Fig. 1(a), dissolved hydrogen prevented a full BCC to HCP plate martensite transformation during quenching and it stabilised some metastable BCC phase at room temperature. In comparison, Fig. 1(b) shows that a full BCC to HCP plate martensite transformation occurred in the non-hydrogenated sample. The martensite transformation is diffusionless, therefore the quenched samples retained the uniform composition of the parent BCC grain as shown by the EDS Al map inserts Figs. 1a-b). Hydrogen tempering at 580 °C decomposed the metastable BCC phase to HCP alpha and FCC or FCT hydride phases (Fig. 1c). Some hydride phase nucleated within the coarse martensite plates to form nanobanded hydrides (NH). The nanobanded hydrides are leaner in aluminium (Al) and are adjacent to Al rich regions (insert Fig. 1c). This indicates that hydrides partitioned Al to form the adjacent Al rich regions, and this resulted in a nanobanded structure that has alternating Al lean/Al rich regions. In comparison, non-hydrogen tempered martensite has an expected uniform distribution of Al between HCP plates. However, it has no nanobands and extensive Al enrichment within HCP plates. The [210] Fourier transforms in Fig. 2 show that the nanobanded region has a HCP crystal structure. The Al rich nanoband in Fig. 2(i) has superlattice reflections (red) which confirm Ti_3Al precipitation. The Al lean nanoband in Fig. 2(iii) unexpectedly has an HCP structure. This indicates that hydrogen inadvertently escaped from the thin sample and consequently transformed the FCC or FCT hydrides to HCP alpha. Dehydrogenating at 675 °C or 750 °C removes hydrogen and decomposes hydrides. However, sufficient diffusion is required to dissolve Ti_3Al in order to control its concomitant embrittlement.

References:
1. Vilane, V.N., *et al.* (2015) Mater. Sci. Forum 828-829, 181.

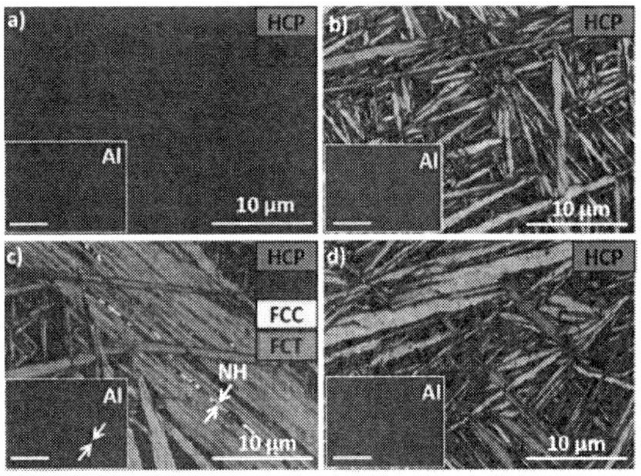

Figure 1. EBSD phase maps of a,c) hydrogenated and b,d) non-hydrogenated martensite in Ti-6Al-4V. Inserts show corresponding EDS Al maps (scale length: 10 µm). Samples were respectively a,b) quenched and c,d) tempered (580 °C). Colour coding in EBSD maps confirms HCP alpha, BCC beta and FCC or FCT hydrides.

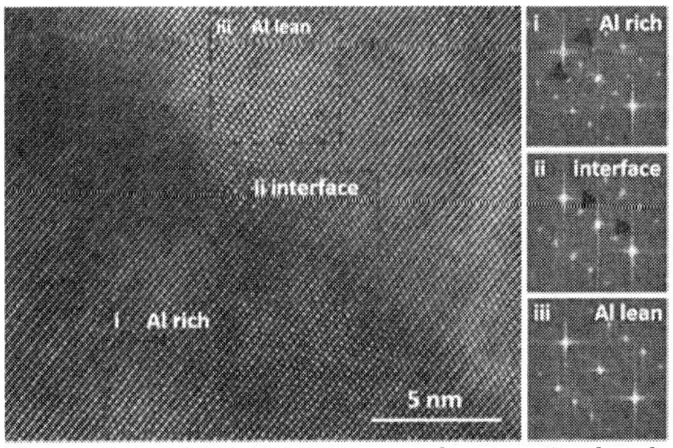

Figure 2. HAADF STEM image showing i) Al rich (dark) and iii) Al lean (bright) nanobands and their ii) interface. The corresponding [210] Fourier transforms confirm an HCP crystal structure at all regions. The Al rich region has superlattice reflections (red) which confirm Ti_3Al, some of which show in ii).

Corresponding author: vlnvel002@myuct.ac.za

MICROTEXTURE DEVELOPMENT IN AISI430 FERRITIC STAINLESS STEEL AFTER HOT ROLLING

R.D. Knutsen, R.R. Masindi and S.L. George

[1]Centre for Materials Engineering, Department of Mechanical Engineering, University of Cape Town

AISI430 ferritic stainless steel is predominantly used in sheet forming and hence formability and surface finish are critically important. The microstructure and crystallographic texture that develops in the steel during thermo-mechanical processing influences both aspects but the present work makes particular reference to surface ridging that is associated with texture inhomogeneity[1,2]. AISI430 commonly occurs as a two-phase microstructure (primary ferrite and austenite) at hot rolling temperatures, and during cooling the austenite will either transform to secondary ferrite or martensite depending on the cooling rate. Figure 1 indicates typical mixed microstructure containing all 3 constituents as a result of the cooling rate being insufficiently slow to allow full transformation of austenite to fine grained secondary ferrite. The residual austenite has transformed to martensite but the primary ferrite (large elongated grains) remain unchanged. Consequently, the deformation textures developed in the respective ferrite and austenite phases during hot rolling, and the transformation textures that evolve during cooling, as well as possible superimposed recrystallization, create significant texture inhomogeneity (microtexture). The nature of the texture inhomogeneity is influenced by the austenite decomposition path and the present study investigates the differences in microtexture evolution during continuous austenite to secondary ferrite transformation versus martensite formation and subsequent tempering.

An AISI430 heat was selected for its high austenite volume fraction at hot rolling temperatures (approx. 30 vol.%). Specimens were deformed by plane strain compression at 900 °C and either transformed isothermally at 700-800 °C (austenite to ferrite transformation) or rapidly cooled to room temperature and reheated to 700-800 °C (martensite tempering). Microtexture was investigated using electron backscattered diffraction (EBSD) in the SEM at 20 kV and beam currents in the range 1-5 nA. The map step-size varied from 0.15-2 mm depending on the required level of resolution and area of interest. The pattern quality (or band contrast) proved useful to illustrate the microstructures (as in Fig. 1 and Fig. 2 (inset)).

Since ferrite (primary and secondary) and tempered martensite have the same BCC lattice, microtexture cannot be distinguished by phase. Instead the orientation data was split according to grain size to distinguish between texture contribution of the primary ferrite and either of the fine grained secondary ferrite or tempered martensite blocks and packets. An example is illustrated in Fig. 2 where the fibre diagram (determined from the texture orientation distribution function) demonstrates a shift in preferred orientation intensity from (111)[-1-12] to (332)[-1-13] as the contribution of tempered martensite is isolated. In this case martensite is mostly represented by grain sizes < 15mm.

Comparison of large area EBSD maps (4 mm^2) for the two austenite decomposition paths indicates stronger {001}<110> and {112}-{113}<110> for the continuous austenite to ferrite transformation. The microtexture analysis highlights the possibility that stronger {001}<110> arises due to ferrite transformed from recrystallized austenite. Austenite recrystallization is likely as a result of the incubation time required for ferrite transformation during the isothermal step. The {112}-{113}<110> can be associated with preferred growth of deformed ferrite as the transformation progresses. For tempered martensite, stronger contribution is provided by {332}<113> orientations (Fig. 2), possibly due to variant selection during the martensite transformation and selective growth during the tempering reaction.

References:
1. Knutsen, R.D. and Wittridge, N.J. (2002) Mat. Sci. Technol. 18, 1279.
2. Engler, O., Huh, M-Y. and Tomé, C.N. (2005) Metall. Mater. Trans. 36A, 3127.

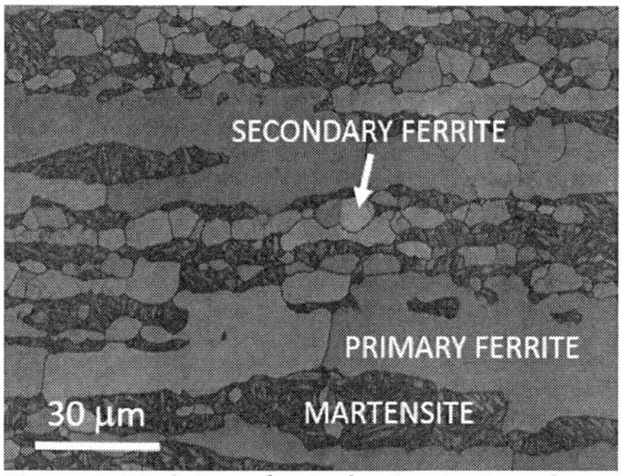

Figure 1. Evolution of mixed microstructure during cooling after hot rolling (EBSD band contrast image).

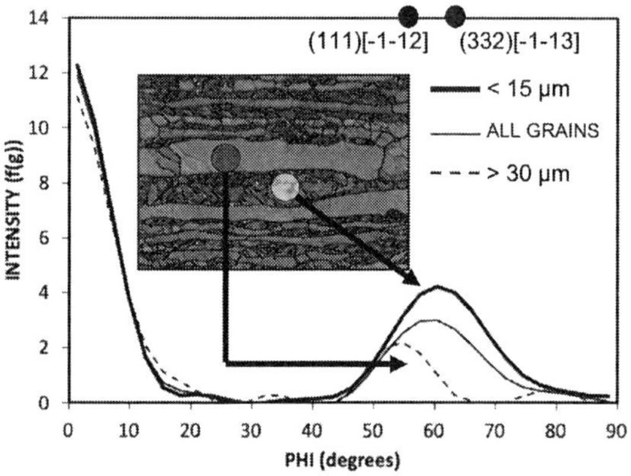

Figure 2. Microtexture after martensite tempering at 700 °C (fibre diagram: Phi1=90°, Phi2=45°).

Corresponding author: Robert.Knutsen@uct.ac.za

MISCROSTRUCTURAL ANALYSIS OF PROMOTED FISCHER- TROPSCH CATALYST ENCAPSULATED INSIDE FUNCTIONALIZED HOLLOW CARBON

T.I. Buthelezi , R.P. Forbes and N.J. Coville

Molecular Sciences Institute, Department of Chemistry, University of the Witwatersrand, Johannesburg

Fischer-Tropsch synthesis (FTS) is the catalytic hydrogenation of carbon monoxide over the surface of a transition metal catalyst to produce hydrocarbons and oxygenates[1]. Compared to transition metals such as Fe and Ru[1,2], Co [2] is a more widely used FTS catalyst due to the high activity of the Co active sites, high selectivity towards linear hydrocarbons, good stability and relatively low cost. One way of enhancing the activity of a Co catalyst is by promotion, i.e. addition of small amounts of a noble metal[3], such as Pt or Ru, to the Co. The noble metal is able to dissociate hydrogen gas at a lower temperatures than the catalyst thus promoting the reduction of cobalt at a lower temperature than in the absence of the promoter. This process is known as hydrogen spillover[3] and is dependent on the type of contact formed between the promoter and the metal oxide. Porous carbon-based supports can also impact on FTS catalysts[3]. Both approaches enhance the stability, dispersion and electronic conductivity of the Co catalyst[2,3].

In this study, Ru (0.5 wt%) promoted and unpromoted Co (10 wt%) catalysts were encapsulated within a porous support, *viz.* nitrogen doped hollow carbon spheres (Co-Ru@NHCS and Co@NHCS), to study the hydrogen spillover process. The catalysts were synthesized by a hard templating method, using polystyrene as the hard template[3]. Catalysts were structurally characterized using SEM, TEM, XPS and BET techniques. The reducibility of the catalysts was assessed using TPR and in-situ PXRD analysis. The activity of the catalysts was tested under low temperature (220 °C) FTS conditions.

The microstructural SEM study showed that a spherical polystyrene template was successfully obtained with good dispersion (Figure 1). The TEM study showed that the NHCSs had a spherical and a hollow morphology (diameter of 360 nm ±20 nm and a shell thickness of 40 ±15 nm; Figure 2), which encapsulated and stabilized the metal nanoparticles within its hollow core. Successful incorporation of N-atoms within the carbon matrix by doping methods was confirmed (XPS analysis) and the N-atoms influenced the surface area of the NHCSs (BET analysis). TPR and situ PXRD studies showed that the degree of reduction of the promoted catalyst was due to the hydrogen spillover effect. However, FTS results showed that the promoted catalyst had lower catalytic activity than an unpromoted catalyst, possibly due to the low mass diffusion of reactants and products.

References:
1. Xiong, H et al. (2014) App. Cat. A: 482, 377
2. Xiong, K. et al. (2010) App. Cat. A. 389, 173.
3. Phaahlamohlaka, T. N. et al. (2017) ACS Cat. 7, 1568.

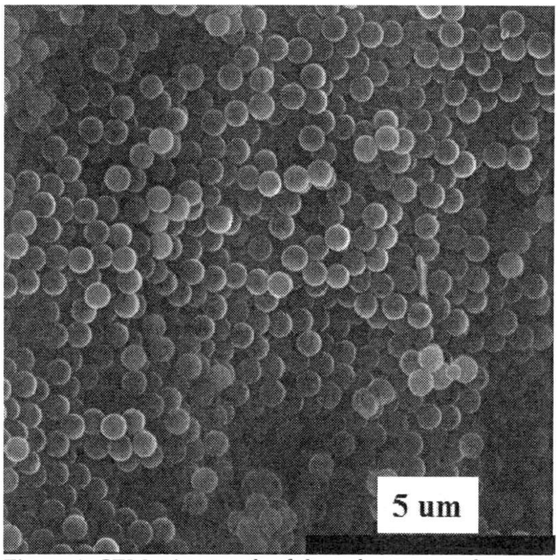

Figure 1. SEM micrograph of the polystyrene spheres.

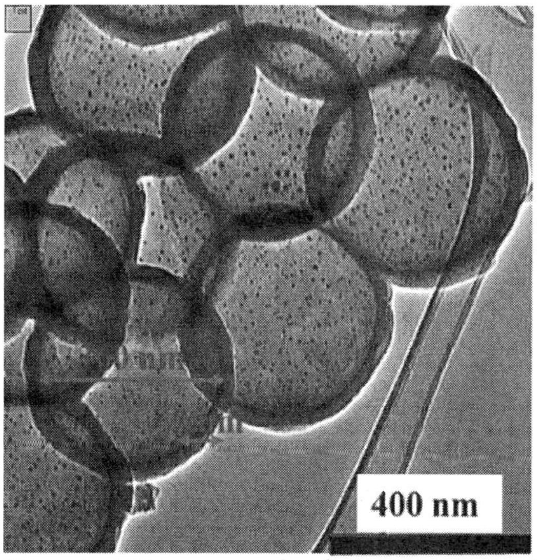

Figure 2. TEM micrograph of metal encapsulated catalyst (Co-Ru@NHCSs).

Corresponding author: neil.coville@wits.ac.za

MISORIENTATION AND GRAIN SIZE EFFECT ON CREEP PROPERTIES OF GRADE P22 STEEL

S. Von Willingh , R.D. Knutsen and R. Curry

EPPEI Materials and Mechanics, Centre for Materials Engineering, Department of Mechanical Engineering, UCT, Cape Town

The study of non-equilibrium steel microstructures, including bainite and martensite, is attracting significant attention in recent times not only because the life cycle management of ferritic steels in ageing power plant is critical, but modern experimental tools such as high speed – high resolution electron backscattered diffraction (EBSD) provide new insights into the behaviour of different microstructures during mechanical loading[1]. Although the true orientation relationships between prior austenite and martensite and/or bainite continues to be debated in the open literature, the different categories of grain boundary within a prior austenite grain (PAG) undoubtedly contribute to creep strength in various ways[2]. Furthermore, the internal grain boundary structure in each PAG is influenced by the PAG size. In the present research, the creep resistance of Grade P22 steel (2¼ Cr) is measured as function of PAG size, where the latter has been tailored through controlled heat treatment at a range in austenitisation temperatures. The research is motivated by the need to gain better insight into the incidence and cause of creep failures in the heat affected zone (HAZ) of thick-walled components on fossil fired power plants.

Creep tests were conducted under constant load (initial stress = 150 MPa) in an air atmosphere at 600 °C. Uniaxial tensile specimens were tested in the solution treated (ST) condition. The solution treatments were performed on the creep test specimens at 900 °C, 1000 °C, 1100 °C and 1200 °C, and soaked for one hour in each case, in order to develop different PAG sizes. The PAG size and analysis of the microstructures were investigated using EBSD in the SEM. EBSD data acquisition was performed using the Oxford Instruments AZTEC system fitted to a TESCAN Mira 3 SEM. EBSD pattern quality was optimized at 20 kV to obtain an indexing rate greater than 80% at beam step intervals of 0.1 μm. Large area maps measuring at least 0.36 mm^2 were acquired from stitching individual 100x100 μm^2 fields.

As anticipated, the PAG size increases as function of ST temperature in the order 22 μm (900 °C), 24 μm (1000 °C), 70 μm (1100 °C) and 108 μm (1200 °C). The creep test results are plotted for the ST condition in Fig. 1 and the PAG size is included on the secondary axis. The time to failure increases up to PAG size equal to 70 μm (1100 °C) and decreases quite dramatically for PAG size equal to 108 μm (1200 °C). There are two unusual observations in these results, namely the marked difference in creep life for the ST900 and ST1000 conditions despite similar grain size, and the rapid drop off in creep life for the ST1200 condition. To assist in interpreting this behaviour, analysis of the microstructure was performed by studying the correlated misorientation angle distribution (CMAD) determined from the large area EBSD maps. The variant selection that occurs during the austenite decomposition, as a result of the fixed orientation relationship between austenite and predominantly bainite microstructure, produces several clearly defined peaks in the CMAD plot. A portion of the CMAD for the ST900 to ST1200 conditions prior to creep testing is presented in Fig. 2. The peaks at approximately 60° show an increase in relative frequency as the ST temperature increases. Since the peak at 60° is by far the strongest peak for the misorientation angle distribution, significant increases in the relative frequency imply increasing dominance of these special block boundaries relative to the PAG boundaries. It is likely that the high angle stable block boundaries for the ST1200 condition resist dislocation movement and promote strain partition to the large continuous PAG boundaries. Consequently, any weakness at the latter boundaries (e.g. segregation) will lead to premature fracture.

References:
1. Morsdorf, L. *et al* (2015) Acta Mater. 95, 366.
2. Payton, E.J. *et al* (2012) Scripta Mater. 66, 1045.

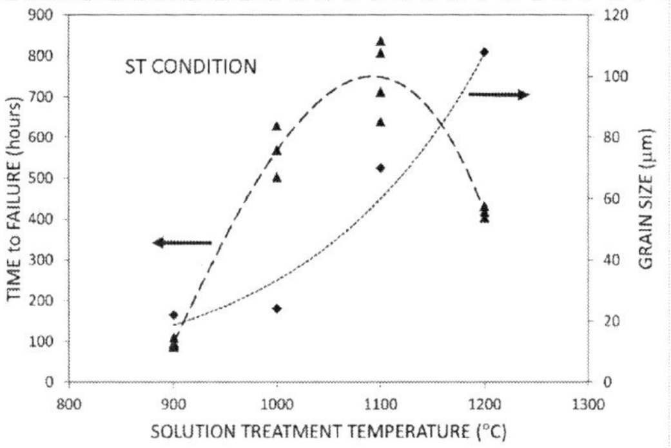

Figure 1. Creep life and grain size for different ST conditions.

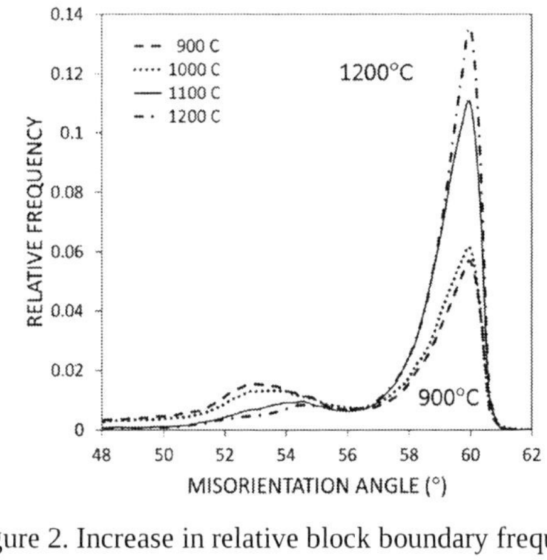

Figure 2. Increase in relative block boundary frequency.

Corresponding author: soraya.vonwillingh@uct.ac.za

MORPHOLOGICAL DEGRADATION AND (BIO)DEGRADATION OF LDPE FILMS TRIGGERED WITH BIOBASED PRO-OXIDANT ADDITIVES

A. Gada[1,2], S. Muniyasamy[1,2], C. Hoyo[3], P. Hlangothi[1], P. Melariri[4] and J.C. Bonner[3]

[1]Department of Chemistry, NMU, PE, [2]Nanostructured and Advanced Materials, Chemicals Cluster, CSIR, Pretoria, [3]Epidemiology and Environmental Epigenetics Lab, NCSU, USA , [4]Department of Environmental Health, NMU, PE

Dating back to the 1950s, the global production of petroleum-based plastics has continued to grow and is predicted to reach 355 million tonnes by 2020[1]. After service life, the post-consumer single use plastic items will accumulate in natural environments if not properly managed as waste, and will persist in the environment to decades and decades of years due to their non-biodegradability[2,3]. This study is aimed at strategically facilitating polyethylene (PE) disintegration and subsequent biodegradation by the introduction of chemical radical reaction initiators and natural polymers in the polymer backbone by use of biobased pro-oxidants.

Two LDPE film formulations with 1% and 3% w/w biobased pro-oxidant additive were prepared by use of a twin-screw melt extruder. The extruded LDPE formulation (1% and 3% pro-oxidant additives) was blown into a 25-27 µm thick film using a melt blower machine. The LDPE (1% and 3%) test films together with PnP LDPE films (control) were submitted to thermal oxidizing test conditions in a 70 °C air ventilated oven and direct sunlight (photo oxidation). The thermally and photo-oxidized LDPE test film were further subjected to biotic degradation tests in aqueous, soil and compost environments. For this study, the surface morphologies of the test samples were recorded using a JEOL-JSM 7001F scanning electron microscope using secondary electron (SE) imaging.

Fig. 1 shows the test samples before abiotic oxidation exposure. Morphological results indicate surface cracks in the samples after 3 months of exposure to direct sunlight and thermal oxidation. After sunlight oxidation degradation exposure, a degree of oxidation increased crack formation on the LDPE-3% film within 113 days whilst PnP and LDPE-1% remained similar even after 200 days of exposure (Fig. 2). A similar pattern was observed in the test samples exposed to thermal oxidation conditions in a 70 °C air ventilated oven after 113 days as compared to LDPE-1% and control LDPE test samples. (Fig. 3). The surface erosion of LDPE additivated samples could be due to the chemical oxidation reactions which was catalysed by UV-light and temperature. The fragmented low molecular weight LDPE films samples were further tested in soil, compost and aqueous media conditions.

Oxidized low molecular weight oxygenated products were shown to undergo assimilation by means of active microorganisms with final by-products of CO_2, H_2O and new microbial cell biomass. The facilitation of LDPE films disintegration and fragmentation by photo-oxidation and thermal oxidation and subsequent biodegradation by use of biobased pro-oxidant additives was achieved.

The authors would like to thank the NRF-DST Waste Research, Development and Innovation (RDI) funding 2017/18, IRG - Egypt / South Africa Research Cooperation Programme 2017/19 and CSIR – Parliamentary grant 2017/18. The authors would like to acknowledgement the Centre for HRTEM for their assistance with SEM imaging. A special acknowledgement to Innoventon-Nelson Mandela University and the CSIR Materials Science and Manufacturing Polymers & Composites unit for assistance with chemical characterizations.

References:
1. Andrady, A.L. and Neal, M.A. (2009) Philosophical Transactions of the Royal Society B: Biological Sciences. 364(1526), 1977.
2. Oehlmann, J. *et al.* (2009) Philosophical Transactions of the Royal Society B: Biological Sciences. 364(1526), 2047.
3. Teuten, E.L. *et al.* (2009) Philosophical Transactions of the Royal Society of London B: Biological Sciences. 364(1526), 2027.

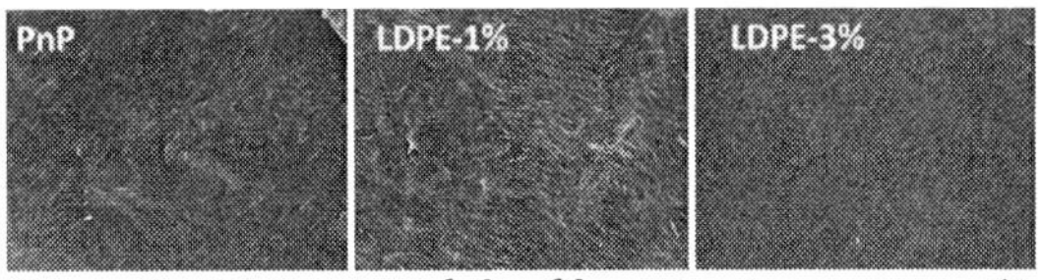

Figure 1. SE images of the film prior to exposure. (1 µm scale)

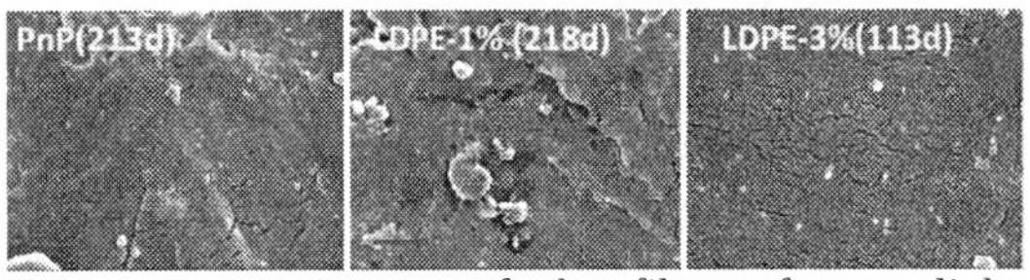

Figure 2. SE images of the films after sunlight exposure. (1 µm scale)

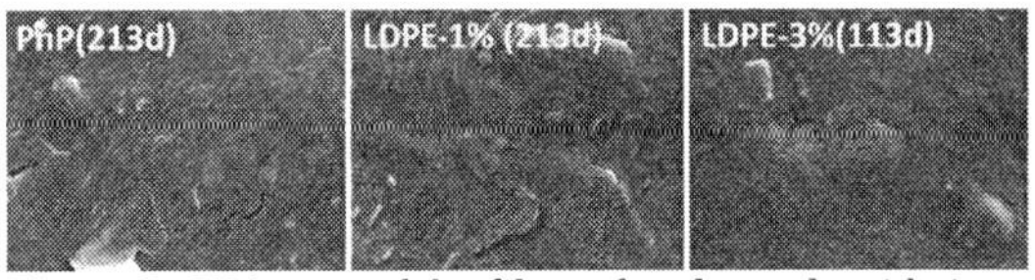

Figure 3. SE image of the films after thermal-oxidations exposure. (1 µm scale)

Corresponding author: abongilegada9@gmail.com

NANOINDENTATION HARDNESS OF TiN NANOPARTICLES STRENGTHENED SAF 2205

M.R. Mphahlele[1], S.R. Oke[1,2], O.O. Ige[1,3], O. Falodun[1], M. Netshia[1] and P.A. Olubambi[1]

[1]Dept. of Engineering Metallurgy, University of Johannesburg, JHB, [2]Dept. of Metallurgical and Materials Engineering, Federal University of Technology, Akure, Nigeria, [3]Dept. of Materials Science and Engineering, Obafemi Awolowo University, Ile-Ife, Nigeria

Recently, researchers are focusing on nanocomposites which are shown to have better hardness and strength than composite materials[1]. In the reported work[2] of the current study, it is demonstrated that duplex stainless steel (SAF 2205) reinforced with 20nm-particles of titanium nitrite (TiN) displays an improvement in the hardness of the steel. In this study, an attempt is made to establish a correlation between TiN content and yield strength of the SAF 2205-TiN nanocomposites.

Materials used in this study were as received spark plasma sintered SAF 2205 reinforced with varying amounts (0-8wt%) of TiN particles 20nm in size. The samples were metallographically prepared to a mirror surface finish using 0.2 µm fumed silica suspension. Nanoindentation tests were performed using ultra nanoindenter (UNHT) equipped with a three-sided Berkovich. A maximum load of 1 mN and holding time of 10 s was used to probe an average of 10 indents per sample. The penetration depths of the indents range from 53 to 55.7 nm, which are higher than the average surface roughness (Ra) of typical surface profiles of polished SAF 2205 (using 2.5 µm particles) which are reported to be <20 nm[3]. The indentation data was then used to estimate the yield stress (σ), according to this relationship[4]: $\sigma = (H_{IT} \times E_s \times h^2_{max})/F_{max}$; where σ is the indentation hardness (GPa), E_s is the sample's elastic modulus (GPa), h_{max} is the maximum penetration depth (mm) and F_{max} is the maximum load (mN).

The scanning electron (SE) micrographs (Fig. 1a) shows the microstructure features of the sintered SAF 2205-6TiN which through the aid of energy dispersive spectroscopy (EDS) analysis (Fig. 1b), the presence of titanium (Ti) along the grain boundary regions is indicated. The presence of Ti at the grain boundaries reinforces the material. This effect is evidenced by an increase of the nanohardness which linearly increased with increasing TiN content. Thus, an empiric correlation was set up between the nanohardness and TiN content. An occurrence of the deviation as noted in figure 2, could be as a result of a combination of influence of pile-up or sink-in effects and indentation size effects. The results of this study show that the hardness (and possibily yield strength) of SAF 2205 can be improved by TiN additions.

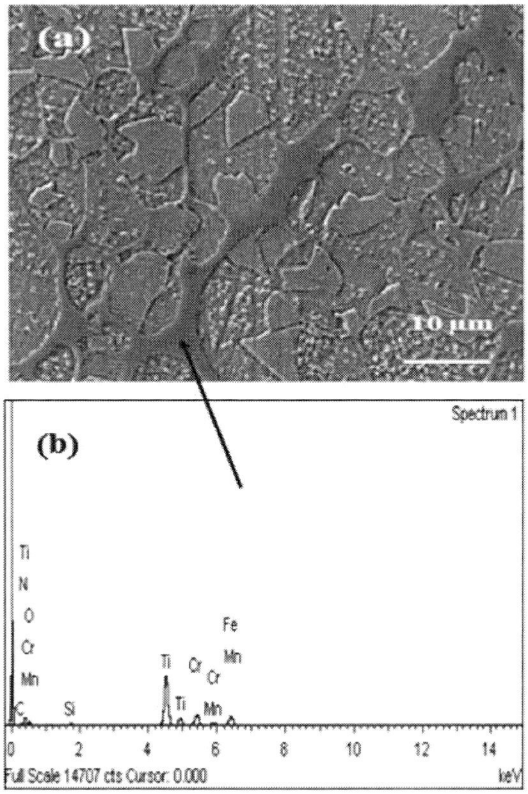

Figure 1. SE micrograph and EDS analysis showing the dominance of Ti on the grain-boundary of SAF 2205-6TiN.

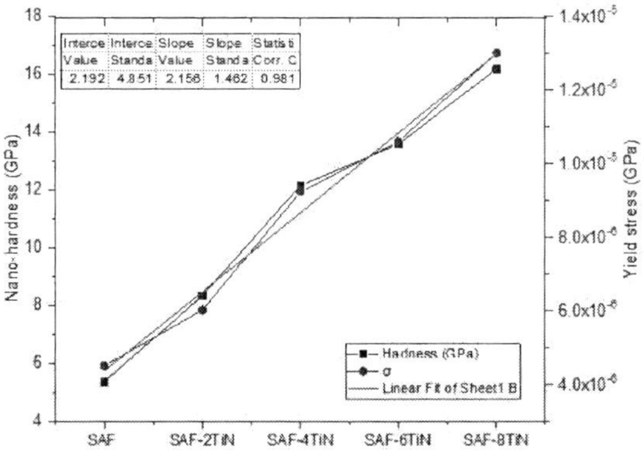

Figure 2. Nano-hardness and yield stress graph of SAF 2205 and its nano-composites.

Corresponding author: ramimphahlele@gmail.com

References:
1. Oke S.R. *et al.* (2018) Proc. Mat. Sci. and Eng., 430, 012034.
2. Mphahlele, M.R. *et al.* (2019) Proc. Mat. Sci. and Eng. 499(1), 012009.
3. Gao, C and Liu, M. (2017) J. of Non-Cryst Solids, 475, 159.
4. Tang Y. *et al.* (2019) J. of Electrochem. Sci., 14, 6794.

NITROGEN-DOPED GRAPHENE QUANTUM DOTS ON FUNCTIONALIZED POLYANILINE NANOFIBERS FOR GAS SENSORS

C.M. Masemola[1], E.C. Linganiso[2], N. Moloto[1] and Z. Tetana[2]

[1]MSI, School of Chemistry, Wits, Johannesburg, [2]MMU, Wits, Johannesburg

Discovery of intrinsically conducting polymers (ICP) such as polythiophene, polypyrrole, and polyaniline (Pani) have introduced unique and important polymer properties in the field of chemistry and physics[1]. These polymers have since been studied for applications in advanced technological devices such as bio/chemical sensors, photovoltaic and anti-corrosive coating agents[2]. This is because polymers in general are easy to synthesize, inexpensive and environmentally friendly[3]. Among these polymers, Pani has received much interest owing to its unique and reversible acid/base doping chemistry which can be tuned for special applications such as gas sensors[4].

However, Pani alone cannot be used as an ideal gas sensor material because of its long gas response times and poor mechanical strength[5]. Researchers are therefore focusing on preparing polymer based composite materials involving the strong carbon nanomaterials such as carbon nanotubes (CNTs), carbon dots (CDs) and graphene quantum dots (GQDs) in order to induce strong mechanical support on the polymers[6]. Numerous reports have shown that a sensor's performance is directly related to porosity and ratio of surface area to volume of sensing material for easy diffusion of gases (sensitivity)[2]. Therefore, an electrospinning technique which produces highly porous and continuous polymeric fibers has been employed to achieve materials with these desired properties[7]. The aim of this work is to functionalize electropun Pani-nanofibers with nitrogen doped graphene quantum dots (N-GQDs) in an attempt to address the aforementioned problems and bring new insights to the polymer-based gas sensors.

For Pani synthesis; aniline monomer was chemically oxidized in acid medium with ammonium persulfate via the rapid mixing method[8] and doped with camphorsulfonic acid (CSA). For electrospinning of Pani; a solution of Pani dissolved in N-methyl-2-pyyrolidone (NMP) was mixed with polyacrylonitrile (PAN) dissolved in dimethylformamide (DMF). Graphene quantum dots were synthesized via a microwave-assisted hydrothermal method (MWHM)[9]. The resultant pale-yellow GQDs solution was precipitated with ethanol solution. The prepared materials were analyzed with transmission electron microscopy (TEM), energy dispersive X-ray spectroscopy (EDS), Fourier transform infrared spectroscopy (FTIR), ultraviolet-visible spectroscopy (UV-vis.), and X-ray powder diffraction (XRD).

The XRD patterns of CSA/Pani showed an intense sharp peak at 2 theta=15^{0} due to CSA group insertion into Pani matrix. This insertion of CSA group was also supported on FTIR spectrum by the reduction of frequencies at 1498 and 1588 cm^{-1} corresponding to benzenoid (IB) and quinoid (IQ) rings due to salt formation. The <R> value (IQ/IB) of CSA/Pani increased to 0.98 resembling an emeraldine salt (ES) of Pani (<R>=1.0 for ES) which can have higher conductivity[10]. The UV-Visible spectrum of CSA/Pani showed three distinct peaks at 330 nm, 450 nm and 630 nm corresponding to the benzenoid ring, CSA group and quinoid ring. TEM images of CSA/Pani showed a semi-flexible rod-like morphology. TEM of GQDs showed a dot-like spherical morphology with dimensions less than 5 nm.

Successful incorporation of CSA group was observed on XRD peak at 2 theta=15^{0}. This CSA insertion also improved the crystallinity of Pani observed by the sharpness of that peak. The value of <R>=0.98 determined from FT-IR suggested a balance between both benzenoid and quinoid rings within the polymer. The GQDs produced via MWHM are well distributed in a narrow range of 1-4 nm corresponding to Figure.1 below.

References:
1. Long, Y. Z. *et al.* (2011) Prog Polym Sci. 36(10), 1415
2. Ogurtsov, N. A. *et al.* (2013) J. Phys. Chem. B. 117(17), 5306
3. Mustapa, R., Abu Mansor, Z. I. and Sambasevam, K. P. (2018) J. Phys. Sci. 29(1), 9
4. Tai, H. *et al.* (2007) Sens. Actuators B Chem. 125(2), 644
5. Deshpande, N. G. *et al.* (2009) Sens. Actuators B Chem. 138, 76
6. Qu, D. *et al.* (2013) Nanoscale. 5, 12272
7. Patil, P. T., Anwane, R. S. and Kondawar, S. B. (2015) Prog Polym Sci. 10, 195
8. Qiang. J. *et al.* (2008) Synth.Met. 158, 544
9. Xiao, Q. (2009) Powder Technol. 189(1), 103
10. Abdiryim, T., Xiao-Gang, Z. and Jamal, R. (2005) Mater. Chem. Phys. 90(3), 36

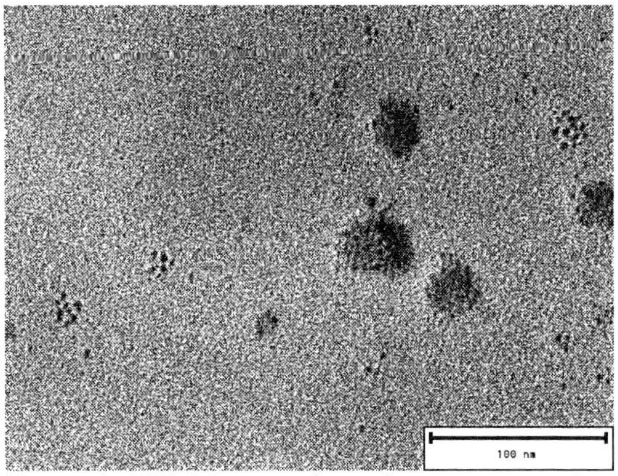

Figure 1. TEM image of graphene quantum dots dispersed in water

Corresponding author: Cebisa.Linganiso@wits.ac.za

OPTIMIZING THE GROWTH CONDITIONS FOR SYNTHESIS OF FEW WALLED CNTS (FWCNTS) USING METHANE

L.R. Mosiane[1,2], B.J. Matsoso[3], N.J. Coville[2], D. Wamwangi[4] and M.S. Maubane-Nkadimeng[5]

[1]DST-NRF Centre of Excellence in Strong Materials, [2]Molecular Sciences Institute, School of Chemistry, University of the Witwatersrand, Johannesburg, [3]Laboratoire des Multimatériaux et Interfaces, CNRS, Université Claude Bernard Lyon 1, Lyon, France, [4]School of Physics, [5]Microscopy and Microanalysis Unit, University of the Witwatersrand, Johannesburg

Carbon nanotubes (CNTs) are amongst the most well known carbon nanomaterials with outstanding mechanical, electrical, magnetic, optical and thermal properties[1]. CNTs have a wide range of potential applications, ranging from biomedical to nanocomposite applications, energy conversion and storage devices to space exploration materials[2]. However, in order for CNTs to attain their potential application in these fields extensive research efforts have been devoted to the tailoring of the physicochemical properties of the CNTs.

In general, CNTs can form spontaneously as either single walled carbon nanotubes (SWCNTs) with one single layer of graphene or as multi-walled carbon nanotubes (MWCNTs) with 2 or more concentric layers of graphites. A special form of MWCNTs, which have attracted interest in the research field, is few walled nanotubes (FWCNTs) with less than 10 layers.

FWCNTs have a semi conducting/metallic behaviour; in this case both the inner and outer layers display unique electronic, optical and Raman scattering characteristics[3]. Furthermore, in contrast to MWCNTs, they possess a higher aspect ratio, enhanced electronic properties and mechanical strength similarl to SWCNTs[4]. Consequently, they could be used in electronic devices to improve the conduction, mechanical strength and light absorption properties of the photovoltaic devices[5].

This study aims to optimize the growth conditions for the synthesis of FWCNT over a Fe-Co catalyst using chemical vapour deposition (CVD). In doing so, the effect of CVD parameters such as flow rate, and the catalyst itself is investigated, in order to determine the role of each parameter on the morphology, nature and structure of the synthesized FWCNTs.

The CNTs used in this study were produced using methane at different hydrogen flow rates over a 10 wt % Fe-Co/$CaCO_3$ catalyst. Transmission electron microscopy (TEM) analysis showed that all the synthesised CNTs were multi-layered and displayed bamboo-like morphology represented by separated inner tubular layers with diameter of 21 ± 5 nm. Raman analysis corroborated the TEM findings; by revealing an increased defect density thus signifying s structural quality deterioration. However, upon change in support material from $CaCO_3$ to MgO, TEM micrographs revealed long range hollow multi layered CNTs with a smaller diameters of 16 nm, with improved thermal stability and yield; confirmed by the TGA data. Additionally, Raman analysis revealed the presence of multiple weak peaks in the frequency range of 100 - 400 cm^{-1}, these are peaks characterised by a radial breathing mode (RBM), observed only for FWCNTs, however this was in contradition with the TEM results. The catalyst loading was then reduced to 5 wt % Fe-Co/MgO whcih lead to the formation of FWCNTs confirmed by both Raman and TEM analysis. Raman data showed strong peaks of the RBM and giving a calculated diameter of the CNTs as 0.9 - 2.3 nm. TEM micrograph further confirms the presences of 2-5 layered CNTs with calculated diameter of 3 ± 5 nm (Fig. 1).

The results of this study show that by adjusting the loading of the catalyst the formation of FWCNTs can be achieved. In future, FWCNTs will be tested in photovoltaic devices as a means to improve the electronic properties of the device.

References:
1. Iijima, S. (1991) Nature 354(6348), 56.
2. Skupov, K. and Adronov, A. (2014) Canadian Journal of Chemistry 92(1),68.
3. Piao, Y. *et al.* (2011) J. Phys. Chem. Lett. 2(13), 1577.
4. Green, A.A. and Hersam, M.C. (2011) Acs. Nano 5(2), 1459.
5. Dillon, A.C. (2010) Chem. Rev. 110(11), 6856.

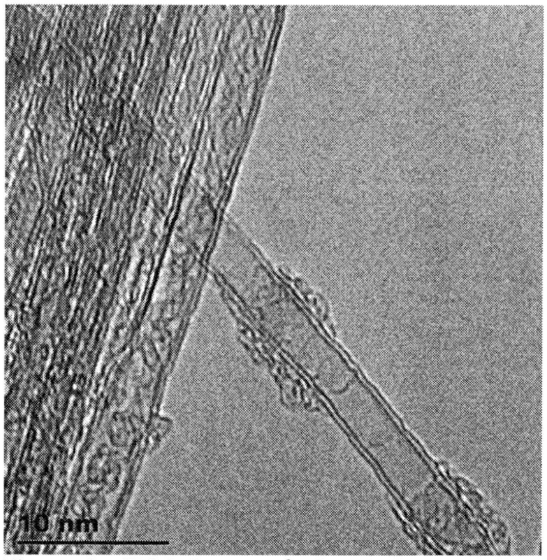

Figure 1. TEM image of CNTs grown on 5% Fe-Co/MgO.

Corresponding author: 840679@students.wits.ac.za

POLYCRYSTALLINE DIAMOND SINTERED WITH ZrB_2 ADDITIVES

M. Jivanji[1,2], R.P. Forbes[3] and J.E. Westraadt[1]

[1]Centre for HRTEM, Nelson Mandela University, Port Elizabeth, [2]Element Six (Pty) Ltd, Springs, [3]Department of Chemistry, University of the Witwatersrand, Johannesburg

Conventional polycrystalline diamond compacts (PDCs) are manufactured by sintering diamond powder on a cemented WC-Co substrate at high-pressure (>5.5 GPa) and high-temperature (≈1450 °C) conditions. During the heating process, cobalt from the substrate melts and infiltrates into the diamond layer to form an interconnected diamond structure. The cobalt remaining after sintering is known to be detrimental as it plays an integral role in initiating catalytic graphitisation above 700 °C[1], leading to catastrophic failure of the PDC tool. The study aims to modify the residual cobalt phase into a non-catalytic phase through ZrB_2 additives.

Diamond powder was mixed with ZrB_2 powder in a 90:10 mass ratio, placed onto a WC-Co substrate, and sintered at 7 GPa and 1750 °C. The sample was investigated using X-Ray diffraction (λ_{Co}=1.789 Å) for phase analysis. Polished cross-sections were imaged using BSE-SEM. Thin lamellae were removed from the binder phases using focused ion-beam (FIB)-SEM. The thin sections were analysed using transmission Kikuchi diffraction (TKD) combined with SEM-EDS and TEM selected area electron diffraction (TEM-SAED) to determine the crystal structure of the binder phases.

XRD analysis (Fig. 1) shows that the sample consists of diamond, WC, ZrC and a $B_6Co_{21}W_2$ (Fm-3m; a_o = 10.46 Å) phase. No free cobalt was detected, although peak overlap complicated the quantification. BSE imaging (Fig. 2) further compliments these results, showing that the overall microstructure consists primarily of polycrystalline diamond particles (dark phase) surrounded by a binder phase. TKD-EDS analysis shows that the binder phase consisted of ZrC and a non-indexed cobalt-rich phase (Fig. 3). Three TEM-SAED patterns were successfully matched with the simulated [110], [213] and [112] zone axis patterns for B_6Co_{23} (Fm-3m; a_o = 10.46 Å).

The results confirm that B from the ZrB_2 additives combined with Co and W to form a $B_6Co_{21}W_2$ phase, while the Zr reacted with the diamond to form large ZrC phases. Ball-milling of the ZrB_2 additive can be optimised to reduce the particle size and prevent the formation of the large ZrC phases. Future work will include investigations of the high-temperature reactions using in-situ hot-stage X-Ray diffraction and vertical turning lath testing to evaluate the efficacy of the ZrB_2 additives on the high-temperature drilling performance of the material.

References:
1. Liu, C. *et al.* (2012) International Journal of Refractory Metals and Hard Materials 31, 187.

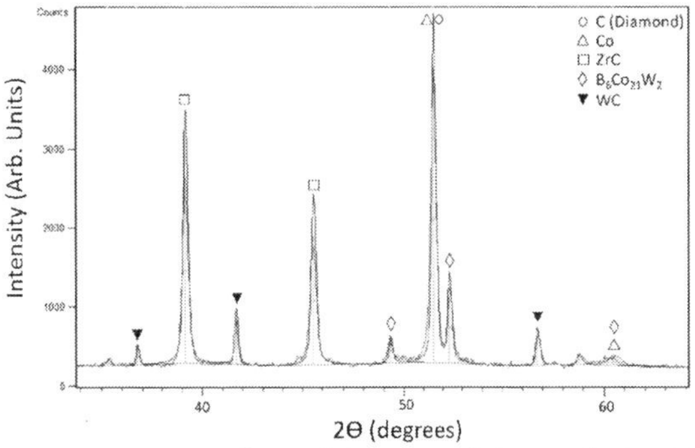

Figure 1. XRD analysis of the sintered diamond-10wt% ZrB_2 powder mixture.

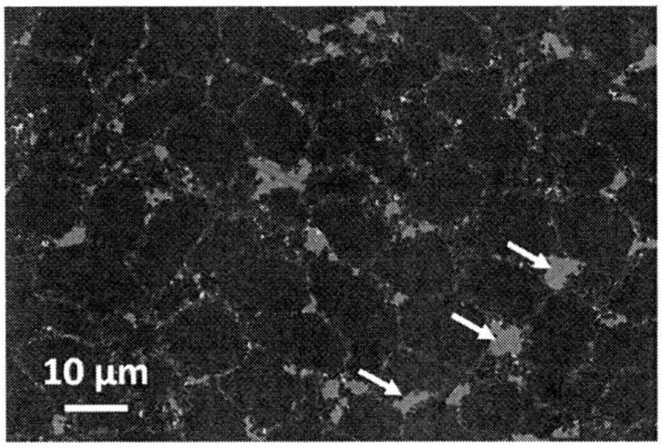

Figure 2. BSE image of the sintered diamond-10wt% ZrB_2 powder mixture. White arrows indicate large binder particles. The red line indicates the position of the FIB lamella.

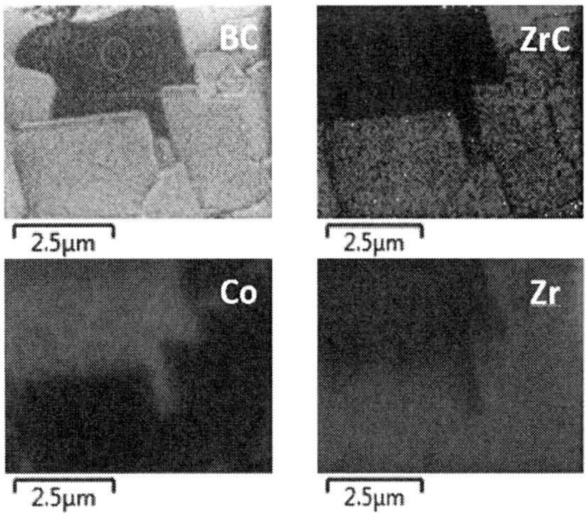

Figure 3. TKD analysis (top) combined with SEM-EDS analysis (bottom) on the thin section removed from the binder pool. TEM-SAD patterns were collected from the unindexed region indicated by the red circle.

Corresponding author: melisha.jivanji@e6.com

POROSITY DEVELOPMENT DURING HYDROGEN SINTERING OF Ti-6Al-4V ALLOY POWDERS

R.D. Knutsen and M.M.S. Mariaye

Centre for Materials Engineering, Department of Mechanical Engineering, University of Cape Town

The sintered density of Ti-6Al-4V alloys can be improved when powder compacts are prepared using TiH_2 (d-phase) powder instead of pure Ti powder[1]. The benefit is attributed to the greater reduction in oxygen content during vacuum sintering of the TiH_2 powders due to the cleansing effect of the dehydrogenation reaction ($TiH_2 \rightarrow Ti+H_2$). In the present study it was postulated that the benefit of using TiH_2 powders could be enhanced by sintering in Ar/15%H_2 gas as opposed to vacuum, and furthermore, that improved sintering efficiency could lead to reduction in sintering temperature. To this end, blended TiH_2 + Al-V master alloy powders were compacted and sintered at 1050 °C for 8 hours in flowing pure Ar and Ar/15%H_2 gases, and in vacuum. The sintered densities are illustrated in Fig. 1 and disappointingly show that the lowest density, even lower than the green density, is obtained for the Ar/15%H_2 gas environment. CP-Ti-6Al-4V is shown for comparison.

The Ti-H phase diagram indicates that TiH_2 should decompose in the following sequence during heating: d→b-Ti→a-Ti. However, the steep concentration gradient caused by flowing Ar will result in local depletion of H from the TiH_2 outer shell leading to stabilization of a-Ti[2]. Given that H solubility in a-Ti (S^{a}_{H}) is lowest (i.e. $S^{a}_{H}<S^{b}_{H}<S^{d}_{H}$), the a-shell will retard further dehydrogenation and possibly favour Al diffusion from the master alloy particles (Al is an a-Ti stabilizer). Consequently, Al enrichment in the a-shell is promoted and V is mostly retained in the master alloy particles. For the flowing Ar/15%H_2 gas case the concentration gradient is much less steep, and it is very likely that there is sufficient H_2 back pressure to stabilize the b-Ti. The b-Ti in the TiH_2 particle outer shell will promote V diffusion from the master alloy and hence causes Al enrichment in the master alloy boundary layer which could potentially promote localized melting above 600 °C. Thus it is feasible to hypothesise that the much lower sintered density for the flowing Ar/15%H_2 gas case is a result of incipient Al melting. To test this hypothesis, the elemental distribution in TiH_2+master alloy sintered in both gas environments was examined in the SEM using energy dispersive X-ray spectroscopy (EDS). Images and X-ray spectra were acquired at 20 kV and probe currents between 0.5-1.5 nA (up to 10k cps).

As anticipated, the Al enrichment in the outer shell of the original TiH_2 particle is very evident in Fig. 2 and Fig. 3 for the flowing Ar gas environment. The residual master alloy particle is enriched in V. However, composition analysis after sintering in flowing Ar/15%H_2 gas demonstrates very similar elemental distribution, and despite extensive exploration, evidence for localized Al melting could not be detected. This situation highlights two possible scenarios: either it is possible that the molten Al is rapidly diffused into the b-shell to transform to an a-shell, and hence appears similar in elemental distribution to that depicted in Fig. 3, or the increase in porosity is caused by internal pressure build-up due to the release of H_2O during sintering. Either way, it is clear that sintering of blended elemental powders containing TiH_2 particles in a H_2-rich atmosphere should be avoided to derive the beneficial cleansing effect as H_2 is liberated during TiH_2 decomposition.

References:
1. Oh, J.M. *et al.* (2013) Materials Transactions 54(1), 119.
2. Jiménez, C. *et al.* (2011) Acta Mater. 59, 6318.

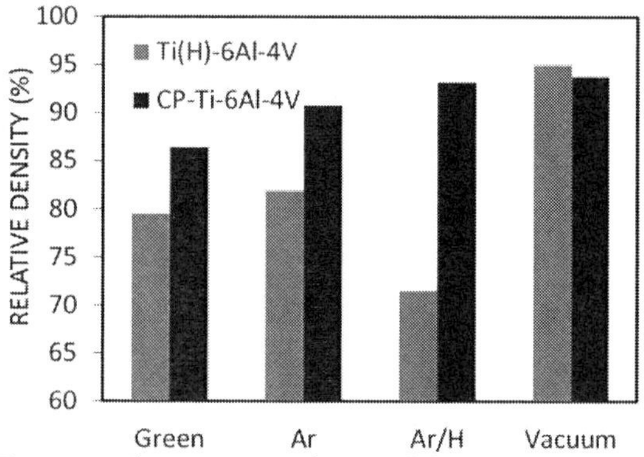

Figure 1. Relative green and sintered densities.

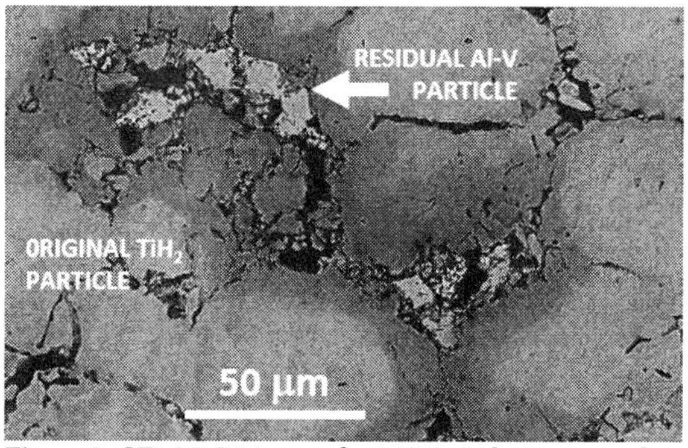

Figure 2. SEM-BSE image for Ti(H)-6Al-4V sintered in flowing argon.

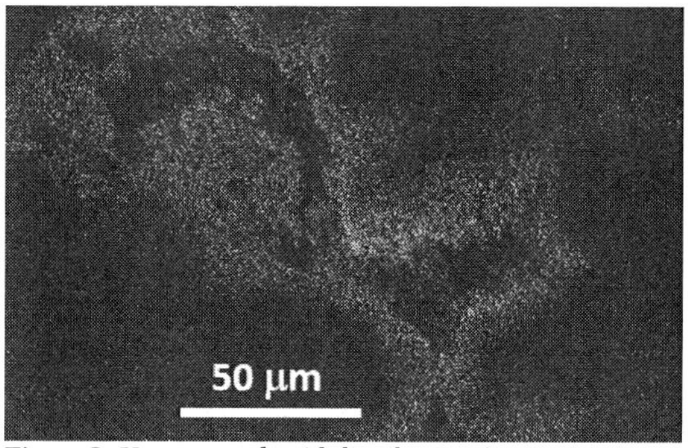

Figure 3. X-ray map for Al distribution (Fig. 2 area).

Corresponding author: Robert.Knutsen@uct.ac.za

PRODUCTION AND PROPERTIES OF A MALEATED CASTOR OIL-POLYSTYRENE POLYMER MATRIX

L. Ferreira and C. Woolard

Center for Materials Engineering, Mechanical Engineering, University of Cape Town, Cape Town

In a world with a rapidly growing population, the usage of plastic products have increasingly negative impacts on the environment. To reduce the negative impact on the environment, biodegradable alternatives for commercial polymeric materials have been considered[1]. Owing to the presence of hydroxyl and carboxyl functionalities, biological alternatives such as castor oil are fit for use in the synthesis of a variety of monomers[2]. These hydroxyl and carboxyl groups have high reactivities [3].

In this study, medical grade castor oil was maleated by the addition of maleic anhydride to form maleated castor oil (MACO). The reaction took place at a temperature of 98°C for 5 hours. The MACO was reacted with styrene monomer in the presence of an initiator (*tert*-butyl peroxybenzoate) at a temperature of 60°C until the reactants were mixed thoroughly. The mixture was cured successively using a vulcanization press for 2 hours at 90°C, 2 hours at 120°C and 1 hour at 160°C. Determination of acid values were used to monitor the maleation of castor oil. The maleated castor oil-polystyrene matrix (MACOPS) was reinforced with greige fibers (55% hemp/45% cotton) using a hand lay-up process. Tensile, flexural and impact tests were conducted to determine the mechanical properties of the synthesized matrix and composite. Raman confocal microscopy (WITec Alpha 300R) at a laser power of 9mW was used to determine the structural properties of MACOPS. Raman mapping was used to study the distribution of polystyrene (PS) through the sample. The WITec RISE (Raman-SEM) microscope was used for fracture surface analysis of the reinforced MACOPS samples.

Raman spectroscopy was used to determine if the maleation of castor oil took place successfully. Maleic anhydride has signature absorption bands at 1850cm^{-1} and 1790cm^{-1}. These peaks were absent in the MACO spectrum, which suggests complete reaction[4]. Signature peaks of both the MACO and PS were present in the spectrum of MACOPS. This may point to an interpenetrating polymer network (IPN) formation where network 1 will be a homo-polymer of styrene and network 2 will be a random co-polymer between MACO and styrene.

A Raman map of MACOPS is seen in Fig. 1 with (a) only showing PS distribution (green) and (b) showing the distribution of MACO (yellow) and PS (green). In Fig. 1 (b) the MACO is seen to overlap some of the green areas in Fig. 1 (a). This could suggest that the MACO chains are connected to PS that forms the co-polymer. The suggested networks are interlaced forming the IPN.

During the mechanical testing of the reinforced MACOPS, a significant decrease in mechanical properties were observed when compared to the MACOPS matrix. This is not expected for reinforced composites. To account for the decrease in properties, the fracture surface was analyzed using SEM. The analysis revealed that the greige fibers consists out of a bundle of smaller fibers. This suggests that the modulus calculated was only for one of the internal small fibers. Voids in the matrix is present, suggesting insufficient wetting of the fibers during hand lay-up. Voids result in a decrease in the mechanical properties of the reinforced MACOPS composite.

MACOPS was determined to be an IPN between PS and MACOPS co-polymer. The poor properties of the reinforced MACOPS composite are due to the presence of voids and weak fiber bundles that only act as a filler.

References:
1. Guner, F.S., Yagci, Y. and Erciyes, A.T. (2006) Progress in Polymer Science. 31, 633.
2. Patel, V.R. *et al.* (2016) Lipid Insights. 9, 1.
3. Totaro, G. *et al.* (2014) European Polymer Journal. 56, 174.
4. Mamat, X., Wang, Y. and Eli, W. (2012) Polymers Advanced Technologies. 23, 1271.

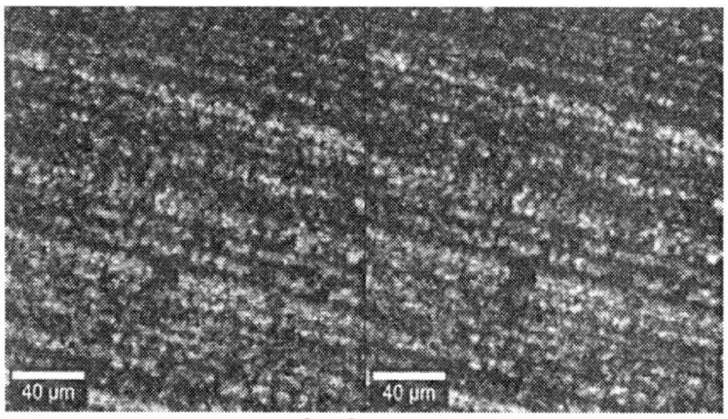

Figure 1. Raman map displaying a) polystyrene (green) and b) MACO (yellow) and polystyrene (green).

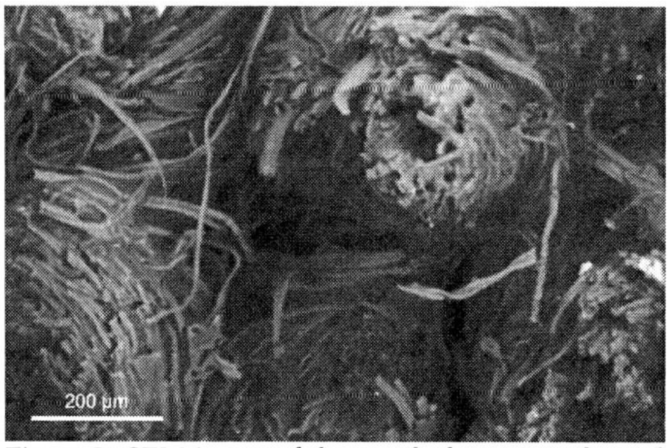

Figure 2. SEM image of the tensile fracture surface of reinforced MACOPS.

Corresponding author: FRRLIZ001@myuct.ac.za

PROPERTIES OF ZIRCONIA REINFORCED Ti6Al4V COMPOSITES PRODUCED VIA SPARK PLASMA SINTERING

L. Semetse[1], B.A. Obadele[1], L. Raganya[1,2] and P.A. Olubambi[1]

[1]Department of Metallurgy, Centre for Nanoengineering and Tribocorrosion, University of Johannesburg, Johannesburg, [2]Light Metals, Materials Science & Manufacturing, Council for Scientific and Industrial Research, Pretoria

Ti6Al4V is an alloy of titanium consisting of 6 wt% aluminum and 4 wt% vanadium. This alloy has outstanding properties like high strength, low density and excellent corrosion resistance, making it a useful material in aircraft parts and implant components. Despite its unique properties, Ti6Al4V has been reported to have poor tribological properties[1]. Various titanium matrix composites (TMCs) have been developed using powder metallurgy techniques and show promise in improving mechanical properties of Ti6Al4V[2].

In this work, spark plasma sintered Ti6Al4V without and with 10 wt% ZrO2 were investigated to determine the effect of ZrO2 on the structural characteristics and fretting corrosion properties. The sintered compacts were prepared using a standard sample preparation procedure, to obtain mirror-like surface finish. Samples were etched in Kroll's reagent (2ml HF + 6 ml HNO3 + distilled water) and their structures examined using Zeiss Supra 55VP scanning electron microscope (SEM) equipped with EDS for composition analysis. For analyzing wear properties, fretting tests were carried out using cylinder-on-plane contact configuration at a load of 85 N for 16 hours, against Ti6Al4V counterpart. Tests were carried out in foetal bovine serum medium. The wear volume and wear scar were analyzed using a Veeco-Wyko NT9100 (Bruker™) optical profilometer.

SEM micrograph of the sintered Ti6Al4V with 10 wt% ZrO2 prior to fretting tests is shown in Figure 1. The structure consisted of hollow and rounded phases (confirmed to be rich in Zr by EDS) dispersed throughout the lamellae alpha and beta matrix of Ti6Al4V. The surface profile of Ti6Al4V after fretting is shown in Figure 2. Considerable wear and material detachment (highlighted by areas in blue) could be observed, indicating higher susceptibility to wear and fretting corrosion. On the other hand, Ti6Al4V with 10 wt% ZrO2 exhibited a lower surface degradation, as shown in Figure 3. The smaller wear track area with minimal material removal under similar testing conditions signifies higher resistance to fretting corrosion. This low material loss can be attributed to the strengthening effect imparted by the ceramic particles in the matrix of Ti6Al4V, thus producing lower wear and least amount of degradation compared to unreinforced Ti6Al4V. Further studies may be necessary on the phase evolution and the effect on other properties.

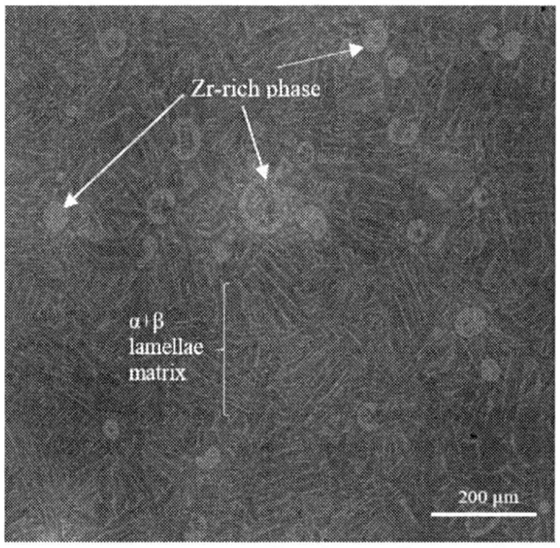

Figure 1. SEM micrograph of Ti6Al4V with 10wt% ZrO_2

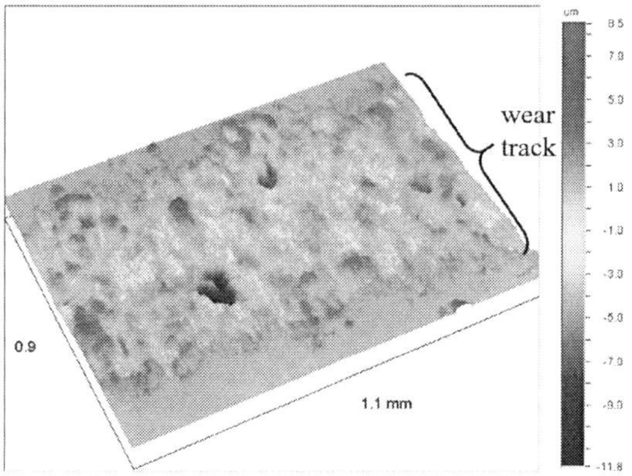

Figure 2. Post-wear optical profile of Ti6Al4V

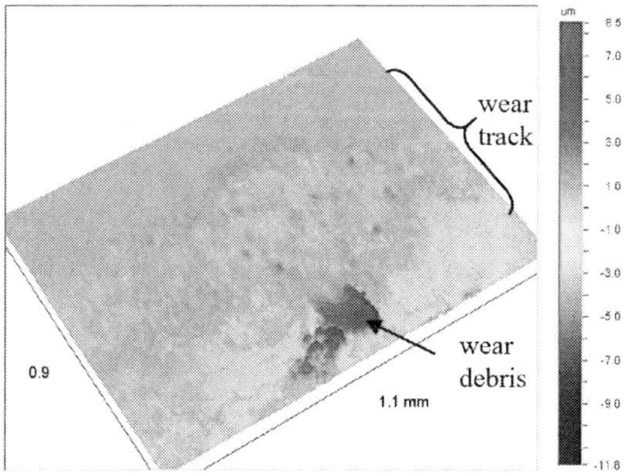

Figure 3. Post-wear optical profile of Ti6Al4V with 10 wt% ZrO_2

References:
1. Leyens, C. and Peters, M. eds. (2003). Titanium and titanium alloys: fundamentals and applications. Weinheim, John Wiley & Sons.
2. Hayat, M.D. et al. (2019) J. Composites Part A. 121, 418.

Corresponding author: lerato.semetse@gmail.com

RAMAN IMAGING AND SCANNING ELECTRON MICROSCOPY (RISE) ON THERMO REGULATING FIBERS

M. Waldron and R. Scheepers

Aaron Klug Centre for Imaging and Analysis, University of Cape Town, Cape Town, [2]Dept. of Textile Science, Nelson Mandela University, Port Elizabeth

Phase change materials (PCMs) are substances which absorb or release large amounts of latent heat when they undergo changes in their physical state. In a heating or a cooling process, this phase change takes place as soon as the material reaches its specific phase change temperature. The PCMs' property of absorbing and releasing large amounts of heat in a controlled way can be utilised to improve the thermal performance of various end-use products to which the PCMs are applied[1]. PCM technology was developed for NASA to protect astronauts from extreme temperatures and consists of small capsules incorporated within the spun fiber, dispersed throughout the fibers in a mat or added as a laminated coating to the fibers. Incorporation within the fiber is only possible with certain types of fiber and dispersion of the PCM capsules throughout the fibers can result in an uneven coating throughout the material. Lamination of the fiber with a thin film of PCM results in the best coverage of the fibers but can alter the flexibility of the textile[2]. This preliminary study was initiated in order to determine a successful and more efficient way to incorporate a PCM into two types of fiber.

Two fiber samples were used in the study: Sample 1 was a combination of nylon, polyester and recycled polyethylene terephthalate, 1 mm thickness and 95 g/m^2 nonwoven sample. Sample 2 was 100% polyester, 15 mm thickness and 120 g/m^2 nonwoven sample. The PCM was a microencapsulated n-paraffin wax with a phase change temperature between 18 °C and 36 °C and two methods of application were used: Pad-dry-cure process and lamination. The fibers that showed successful incorporation of the PCM under the light microscope were examined with Raman Imaging and Scanning Electron (RISE) microscopy. The RISE system operates on a Tescan MIRA3 SEM with a WiTec Alpha 300 Raman system mounted above the SEM chamber. The samples were viewed in low vacuum conditions using the backscatter detector. Images of the fibers were taken with the SEM and then the sample was moved from under the SEM pole piece to the Raman objective in the SEM chamber and a Raman map was acquired of the same area.

Sample 1 showed that lamination of the PCM was successful and an irregular mat could be seen between the fibers (Fig. 1A). Lamination was not successful on sample 2, but the pad-dry-cure coating method appeared to be effective (Fig. 1B). The same area of each sample was then examined with the WiTec Raman microscope and a map collected of the areas of interest. The maps were then overlaid onto the SEM images to clarify the results (Fig. 2A and Fig.2B).

The SEM images of sample 1 show the laminated mat, but lack details about the structure and composition of the mat. However, the RISE results (Fig. 2A) clearly show that the mat is composed of bubbles of PCM suspended within the mat. The SEM images of sample 2 allow us to see bulges on the fibers and RISE information (Fig. 2B) shows clearly that the bulges on the fibers are the PCM coated onto the fiber in an uneven film.

This preliminary study indicated that the PCM microcapsules could be incorporated into sample 1 through lamination. In sample 2, RISE revealed that the incorporation of the PCM is not inside the fiber but is wrapped around the fiber so integration of PCM in this sample through the pad-dry-cure process method was successful. The study indicates how RISE can give additional information about the PCM coating on the fibers in order to establish an efficient method to assimilate PCMs into textiles.

References:
1. Pielichowski, K. and Pielichowski, K. (2014) Progress in Materials Science 65, 67.
2. Lipol, L. (2015) Textile Today 12, 1.

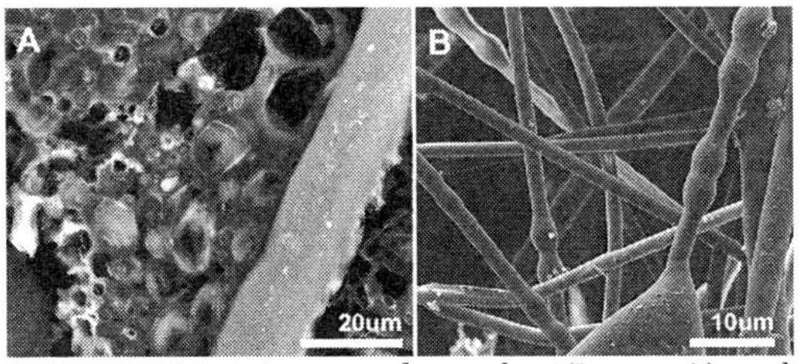

Figure 1. SEM images of sample 1(Image A) and sample 2 (Image B)

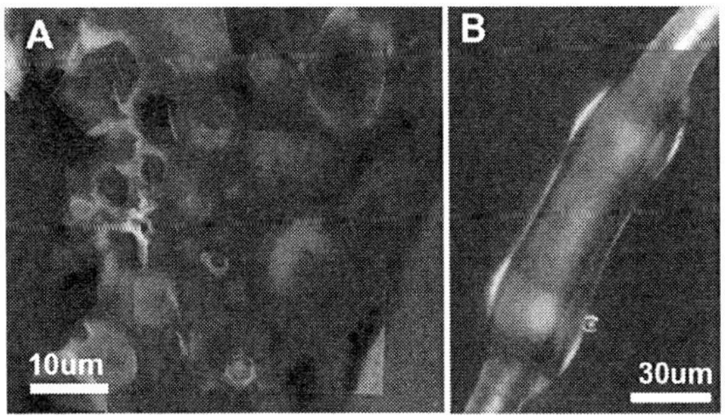

Figure 2. RISE images of sample 1 (image A) and sample 2 (image B) showing the Raman map superimposed on the SEM image

Corresponding author: Miranda.waldron@uct.ac.za

REFACETTING OF PLATINUM-RHENIUM NANOPARTICLES UNDER OPERANDO WATER GAS SHIFT CONDITIONS

R.B. Martin and P.J. Kooyman

Centre for Catalysis, Department of Chemical Engineering, University of Cape Town, Cape Town

The water gas shift (WGS) reaction reacts carbon monoxide and water to form carbon dioxide and hydrogen. This is used in fuel processing for hydrogen fuel cells to reduce the amount of carbon monoxide present in the hydrogen feed line. Carbon monoxide poisons the platinum catalyst used in hydrogen fuel cells[1], therefore it is important that the majority of the carbon monoxide be removed before the feed stream enters the fuel cell so that the fuel cell has a longer lifespan.

Platinum supported on various metal oxides has been identified as a potential catalyst for the WGS reaction, offering higher activity and stability compared to the traditional copper and iron two-step system usually employed industrially[1,2,3].

The changes to the metal nanoparticles during the catalytic reaction have been be studied. The study shows changes in morphology and/or structure, and it also determines whether the particles sinter under reaction conditions. These types of changes affect the catalytic activity and stability, and are of great importance.

Platinum and platinum-rhenium nanoparticles were synthesised by the reduction of $Pt(acac)_2$ and $Re_2(CO)_{10}$, using $W(CO)_6$ as a reducing agent, suspended in dibenzyl ether at 160 ºC for 10 minutes. Oleic acid and oleylamine were used as surfactants. The in-situ studies were carried out in a JEOL 2100 200 kV STEM using a modified Protochips Atmosphere system that allowed water to be introduced to the nanoreactor in a controlled manner. The in-situ studies were carried out at a total pressure between 42 and 50 Torr in an atmosphere of 5% CO, 10% CO_2, 33% H_2O, 30% H_2, and 22% N_2, at temperatures between 320 and 350 ºC.

Figure 1 shows the oscillatory behaviour of a platinum nanoparticle under these reaction conditions. As time progresses, the particle changes shape from one with faceted surfaces (at 0 s), to one which is more spherical (at 7 s), and back again (at 32 s). This behaviour was observed on multiple particles and at different temperatures.

The refacetting is similar to what was observed by Vendelbo *et al.*[4], where Pt nanoparticles were exposed to a mixture of carbon monoxide and oxygen only, at a higher pressure and temperature. This result was not expected as the gas mixture used was very different to the one used by Vendelbo *et al.* The refacetting behaviour is not observed when the samples are imaged under high vacuum in an electron beam. Therefore, this behaviour is a direct result of the reaction gas being present and is similar to what happens in a large-scale reactor.

Acknowledgements
Financial support from the University of Cape Town and the DST-NRF Centre of Excellence in Catalysis (c*change) is gratefully acknowledged. This work is based on the research supported in part by the National Research Foundation of South Africa Grant number 94878. The assistance of Dr Kinga Unocic and use of equipment at Oak Ridge National Laboratory, TN, USA is also gratefully acknowledged.

References:
1. Sato, Y., *et al.* (2005) Applied Catalysis A: General 296, 80.
2. Sato, Y., *et al.* (2006) Catalysis Communications 7, 91.
3. Azzam, K., *et al.* (2008) Applied Catalysis B: Environmental 80, 129.
4. Vendelbo, S.B., *et al.* (2014) Nature Materials 13, 884.

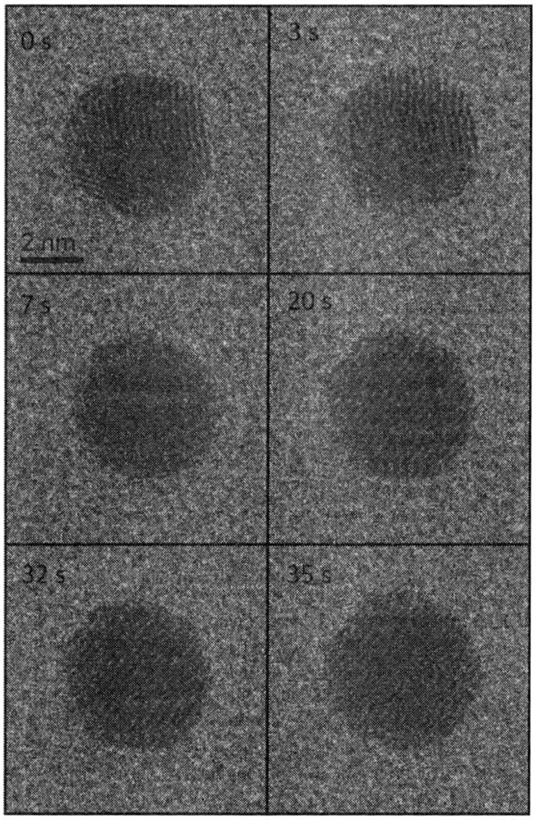

Figure 1. Image series showing a Pt nanoparticle undergoing morphological changes in-situ, 320 °C, 42 Torr.

Corresponding author: mrtric007@myuct.ac.za

SELECTIVE CARBIDE ETCHING OF VANADIUM AND NIOBIUM ALLOYED 25WT% CHROMIUM WHITE CAST IRON

L.A. Mampuru[1,2], L.H. Chown[2] and A. Kapito[1]

[1]Advanced Materials Division, Mintek, Randburg, [2]Sch. Chemical and Metallurgical Eng., and DST-NRF CoE Strong Materials, hosted by UW, JHB

High chromium (25 wt% Cr) white cast iron (HCWCI) is used in the mining industry as liners in mills that crush various ores. Its selection is mainly due to its high abrasion wear and corrosion resistance, attributed to the high amount of hard eutectic and precipitated carbides[1,2]. The type, size and distribution of the carbides contribute to these properties. A typical microstructure expected in HCWCI is a network of M_7C_3 carbides in a mixture of austenite and its transformation products[3].

Failure analysis and metallurgical characterisation have been done on broken parts of milling and mineral ore processing equipment from a range of companies in the mining industry. A recurring problem is breaking, cracking and high wear rates of different mill liners, including those made of HCWCI. These failures are expensive, cause mill down time and unexpected relining and maintenance costs. This prompted the search for ways to extend HCWCI liner life. As the carbide forming elements vanadium and niobium are known to refine carbides, they were added to the HCWCI. Niobium changes the morphology of eutectic carbides. Their morphology is highly dependent on the amount and shape of the austenitic dendrites formed earlier in the solidification sequence[4,5]. A small amount of vanadium does not form VC carbides but reduces formation of needle-like carbides in HCWCI[6]. The main aim is to strengthen and harden the HCWCI by adding carbide forming and refining V or Nb. One objective was to assess the ability of various etchants to clearly reveal the carbides in visible light microscopy (VLM).

As-cast HCWCI and HCWCI +V or +Nb were metallographically prepared for etching to identify the carbides. Various selective carbide etchants were used to reveal the different types of carbides formed in the Nb and V alloyed HCWCI. The microstructures of the etched alloys were studied by VLM and SEM. The morphology, type and distribution of the carbides and the effect of the etchants on microstructure e.g. how carbides were coloured or outlined, were assessed. An visible light micrograph of HCWCI+V in Fig. 1 shows the selective colouring of M_7C_3 carbides by Groesbeck's etchant (100ml water, 4g NaOH, 4g $KMnO_4$) in an austenite matrix. When HCWCI and HCWCI+Nb was etched by 10% ammonium persulfate, known as APS (100ml water, 10g $(NH_4)_2S_2O_8$), the M_7C_3 carbides were outlined, and very fine precipitates were seen in the austenite matrix (Fig. 2 (a) and (b)). Other etchants that were used include alkaline sodium picrate and Murakami's reagent. Overall, the etchants did selectively outline and colour the carbides differently, but all clearly showed the carbide distribution in the alloys.

References:
1. Abd El-Aziz, K. *et al.* (2015) J. Bio. Tribo. 1(25) 1.
2. Karantzalis, A.E., Lekatou, A. and Mavros, H. (2009) JMEP 18(2), 174.
3. Haung, X. (2001) JMEP 10(5) 592.
4. Xiaohui, Z. *et al.* (2008) J. Mater. Lett. 62(6) 857.
5. Fiset, M. *et al.* (1993) J. Mater. Sci. Lett. 12(9) 615.
6. Nurjaman, F. *et al.* (2015) ISFAP 1711, 020005-1.

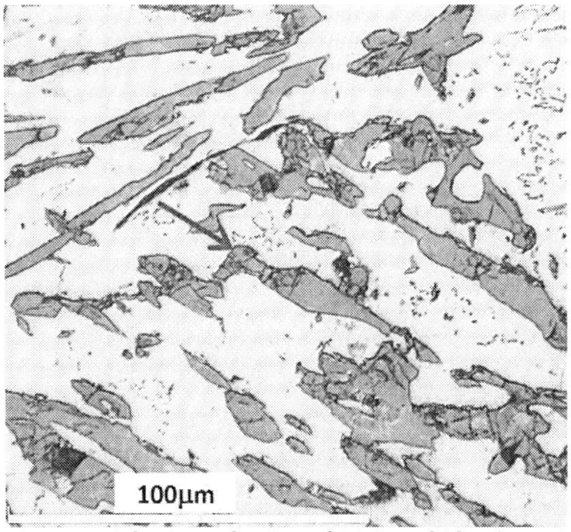

Figure 1. Selective colouring of M_7C_3 carbides (arrow) in a matrix of austenite (white) in HCWCI+V using Groesbeck's etchant.

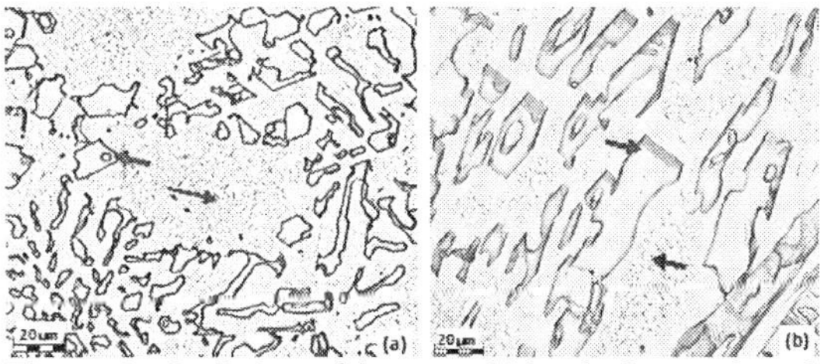

Figure 2. Outlining of large carbides (red arrow) and fine M_7C_3 (blue) in (a) HCWCI and (b) HCWCI+Nb using 10% APS.

Corresponding author: lebidikem@mintek.co.za

SYNTHESIS AND APPLICATION OF BORON AND NITROGEN DOPED CARBON NANO-ONIONS FOR ENERGY STORAGE

B. Shaku[1,2], N.J. Coville[1,3], K.I. Ozoemena[1,3] and M.S. Maubane-Nkadimeng[1,3,2]

[1]DST-NRF Centre of Excellence in Strong Materials , Johannesburg, [2]Microscopy and Microanalysis Unit, University of the Witwatersrand, Johannesburg, [3]Molecular Sciences Institute, School of Chemistry, University of the Witwatersrand, Johannesburg

Different forms of carbon nanostructures have been used in electrochemical energy storage[1], biomedical imaging[2], water treatment[3] etc. Hence it necessitates that carbon materials have to be prepared on a large scale when realizing these interesting applications. One interesting class of carbon nanostructures are carbon nano-onions (CNOs). CNOs have unique physical properties and consist of a concentric layered structure carbon shells that resemble an onion, hence the origin of its name. In the case of energy storage applications in supercapacitors a conductive material that has a high surface area is required. The conductivity of the carbon nanostructures, including CNOs, can be improved by using dopants. Among various dopants, boron and nitrogen are suitable for improving the conductivity of CNOs. The doping process results in good electronic properties of CNOs[4,5].

In this study, doped CNOs were synthesized using a low-cost flame pyrolysis method using grapeseed oil as a carbon precursor. The CNOs synthesized had a particle size less than 100 nm. The CNOs were characterized using transmission electron microscopy (TEM), Thermal gravimetric analysis (TGA) and Raman spectroscopy.

TEM analysis revealed that microstructural features of the as-prepared CNOs from flame pyrolysis were made of particles with nearly spherical shape and are agglomerated in an interconnected network as shown in Figure 1. The doping process did not disrupt the structure of CNOs as confirmed by TEM imaging. Raman analyses show that the introduction of dopants introduces structural defects. Thermal gravimetric analyses show that the materials are not purely carbon due to the presence of boron and nitrogen. Boron and nitrogen were found to modify the specific capacitance of the pristine CNOs at a current density of 0.1 A/g. The amount of energy stored using the doped materials improved as compared to the pristine CNOs. Also, the nitrogen-doped CNOs showed good cyclic stability which indicates that the material can be used as a high power energy storage material. The doped materials also showed improved conductivity as determined from electrochemical impedance spectroscopy. In conclusion, the electrochemical properties of the pristine, boron-doped and nitrogen-doped were investigated using 2 M KOH electrolyte. The experiment done on the full cell showed that nitrogen-doped CNOs performed better in terms of specific capacitance and rate capacibility as compared to the other materials.

References:

1. Jose-Yacaman, M.H. Terrones, H., Rendon, L. and Domingues, J.M. (1995) J. Carbon 33, 669.
2. Sano, N. et al. (2001) J. Nature 414, 5063.
3. Mohapatra, D., Subramanya, B. and Parida, S. (2016) J. Mater. Chem. Phys. 174, 112.
4. Rani, P. and Jindal, V.K. (2013) RSC Advances 3, 802.
5. Panchakarla, L.S., et al. (2009) J. Advance Materials 46, 4726.

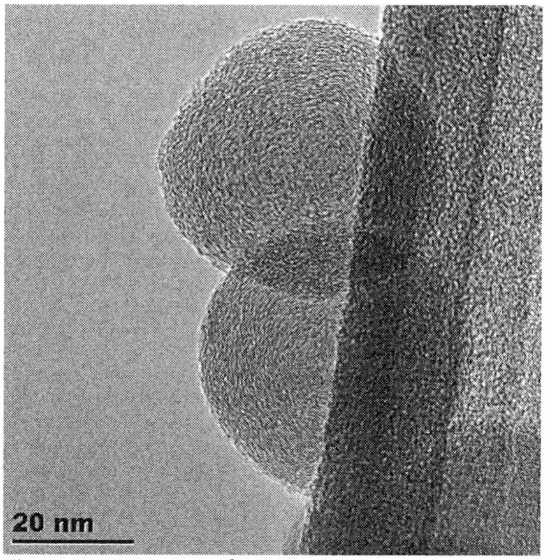

Figure 1. TEM image of pristine CNOs

Corresponding author: 726829@students.wits.ac.za

SYNTHESIS AND CHARACTERIZATION OF DYSPROSIUM (Dy^{3+}) IONS DOPED ZINC SILICATE NANOPHOSPHORS FOR SOLID STATE LIGHTING APPLICATION

P.S. Mbule[1], B.M. Mothudi[1] and M.S. Dhlamini[1]

[1]Department of Physics, CSET, University of South Africa, Johannesburg

Rare-earth ions doped inorganic nanophosphors are attractive due to their nature of exceptional luminescence properties[1]. Researchers have demonstrated that the luminescence properties of nano-scale materials are completely different from their micro or bulk counterparts. This depends on size, shape, morphology, composition and crystallinity of the material. Solid state lighting (SSL) based devices are predicted to play a crucial role in saving a huge amounts of electrical energy and reducing carbon emissions globally. Currently, the research on efficient and low-cost nanophosphors preparation and their use in SSL industries is essential. The dense particle nanophosphors with spherical morphology can increase the screen brightness and improve the resolution because of lower scattering of evolved light and higher packing densities than irregular particles obtained by other synthesis routes[2]. This has paved the way for development of White LEDs, which have an edge over traditional lighting sources due to their compact size, higher transparency and an extremely long lifetime. In this work, morphology and optical properties of dysprosium doped zinc silicate nanophosphors are demonstrated. We investigate their viability for SSL devices.

$Zn_{2-x}SiO_4:Dy_x$(x = 0.5 – 5 mol%) phosphors were prepared using the sol-gel method. 10 ml Tetraethylorthosilicate (TEOS) (Aldrich, 98 %), 3g zinc nitrate hexahydrate (Aldrich, 99 %) and 0.071g dysprosium (iii) nitrate hydrate were used as precursors while ethanol was used as a solvent. TEOS was mixed with ethanol and water, the mixture was stirred for 10-15 minutes. Zinc nitrate hexahydrate was dissolved with 10 ml ethanol and stirred for 10-15 minutes. TEOS solution was slowly added to the zinc solution. Furthermore, 0.5 ml of HCl was added and the solution was subjected to the ultrasonication for 3 hours. $Zn_2SiO_4:Dy^{3+}$ nanophosphor powders were obtained after drying in air. The thermal treatment was carried out in a pre-heated muffle furnace at 1000 °C for 2 hours.

Figure 1 shows the XRD patterns of $Zn_2SiO_4:Dy^{3+}$ nanophosphor annealed at 1000 °C for 2 hours. All the diffraction peaks are consistent with the standard pattern of α-phase Zn_2SiO_4 (card no. 9014832). Figure 2 shows the FE-SEM image of un-doped and $Zn_2SiO_4:Dy^{3+}$ nanophosphor. The samples shows dispersed micro-grain aggregates with irregular and spherical shape. Upon annealing, much smaller nanoparticles with a porous surface morphology were observed. Elemental composition of nanophosphors was confirmed by EDS.

$Zn_2SiO_4:Dy^{3+}$ nanophosphors were prepared by sol-gel method. Subsequently, XRD, FESEM and PL studies were performed to evaluate their viability in solid state lighting applications. XRD results revealed that nanophosphors crystallized to form α-Zn2SiO4 phase after annealing at 1000 °C for 2hours. FE-SEM images showed dispersed micro-grain aggregates with irregular and spherical shape. Upon annealing the sample the nanoparticles appeared to be smaller in size. The relatively intense emission peak shown by PL was observed with the doping concentration of 2 mol%. The intense emission was due to the location of Dy3+ ions at high symmetry site in Zn_2SiO_4. The experimental results suggested that dysprosium doped zinc silicate nanophosphor have a potential use in solid state lighting displays for white light LEDs.

References:

1. Basavaraj, R.B. et al. (2017) Ultrasonics Sonochemistry 34,700

2. Mbule, P.S., Mothudi, B.M. and Dhlamini, M.S. (2017) J. Lumin.192,853

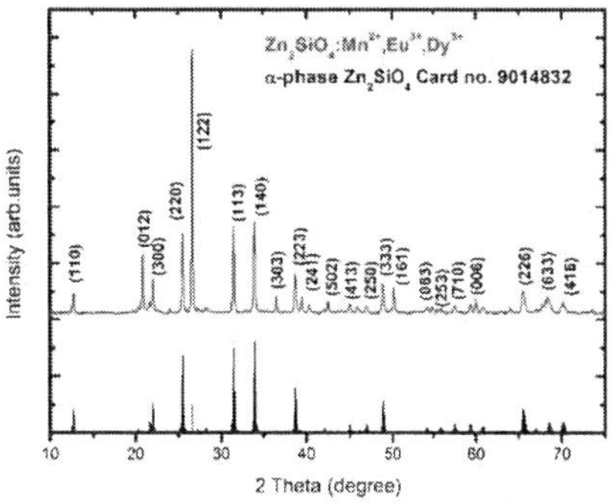

Figure 1. XRD pattern of $Zn_2SiO_4:Dy^{3+}$ annealed at 1000 °C and standard data corresponding to α-phase Zn_2SiO_4.

Figure 2. FE-SEM image of as-prepared Zn_2SiO_4 nanophosphor.

Corresponding author: mbuleps1@unisa.ac.za

SYNTHESIS AND CHARACTERIZATION OF MESOPOROUS TiO_2 NANOPARTICLES FOR THIN FILM PEROVSKITE SOLAR CELLS

D.G. Tsotetsi[1] and P.S. Mbule[1]

[1]Department of Physics, CSET, University of South Africa, Pretoria

Mesoporous Titanium dioxide (mp-TiO_2) is reported to be nontoxic and low cost material that attracted attention from various researchers worldwide due to its potential properties such as high adsorption ability and good photocatalytic activity energy generation[1]. Its properties strongly depend on factors such as large pore diameters (2–50 nm), large pore volumes, surface area, morphology, particle size, crystallinity and ordered or disordered pore channel arrangements[2]. Mp-TiO_2 is considered to be the most used electron transporting material in perovskite solar cells (PSCs) as scaffold and also electron pathway. Highly porous TiO_2 films promote an easy infiltration of perovskite material which subsequently fills the pores. Hence in this study mp-TiO_2 nanoparticles were prepared by sol-gel method with addition of polyethylene glycol (PEG) as co-polymer to control nanoparticle size and morphology. We demonstrate and evaluate the structural, morphological and optical properties of mp-TiO_2 calcined at different temperatures for possible application in thin film perovskite solar cells.

The method for preparing mp-TiO_2 was adopted from Hong et al.[3]. PEG was mixed with 1-propanol and stirred for 1 hour. Solution of deionized water and HCL was mixed to a transparent 1-propanol solution containing PEG and stirred for 10 min. Lastly, TTIP was added to the mixed solution and stirred for 21 hours. After 5 days the completed sol was coated on a glass substrate. The final product had composition of TTIP::PEG::HCL::1-propanol=1::0 0.02::3.2::13.2. The samples were calcined with different temperature of 350°C, 450°C and 550°C. XRD, SEM, UV-VIS and FTIR-RAMAN were used for characterization.

Figure 1 shows the X-ray diffraction (XRD) patterns and phases of the mesoporous TiO_2 nanoparticle prepared by sol-gel method and are calcined at different temperatures. The pattern of the samples presented alike diffraction peaks showing that samples were the TiO_2 nanoparticles consisting of anatase phase confirmed at peak ~25.4° and a rutile phase at peak ~27°. The as-prepared sample showed amorphous-like structure with a rutile phase. The crystallite size depend on the annealing temperature and was estimated to be about 1.048±0.04 nm for as-prepared sample and 17.319±0.04 nm for the sample calcined at 550°C. Figure 2(a-d) shows SEM images of as-prepared mesoporous TiO_2 samples and samples calcined at 350°C, 450°C and 550°C temperatures. The morphology of the samples showed a porous spherical shape and the pores occupied space between the nanoparticles showing that the material is mesoporous.

The Mesoporous TiO_2 nanoparticles were successfully synthesized by the sol-gel method with PEG as a surfactant. The XRD studies showed a mixture of rutile and anatase phases and samples were highly crystalline at ~27° diffraction peak that defined rutile phase after calcining. The morphological results illustrated that the synthesized nanoparticles are spherical in shape due to sol-gel method and the PEG surfactant kept the nanoparticles size uniform. The calcination of nanoparticles at 550°C optimized the porosity of TiO_2.

References:
1. Liu Y.-C *et. al* (2011) Int. J. Photoenergy 2011, 1.
2. Mahoney, L. and Koodali, R.T, (2014), Materials 7, 2697.
3. M.-H. Hong et al.,(2013) Surf.Coat.Tech, 231, 370.

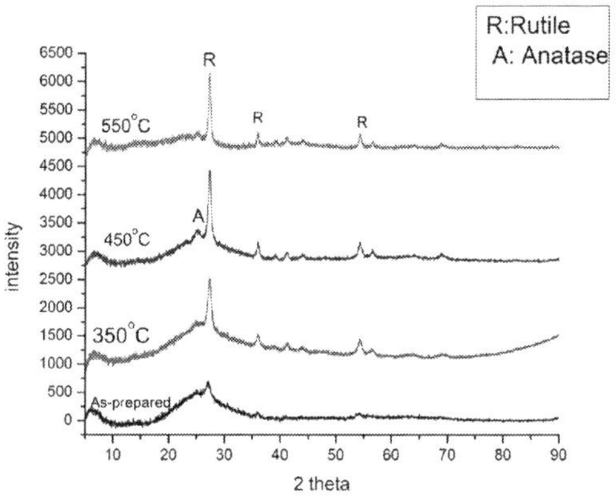

Figure 1. XRD pattern of as-prepared mp-TiO_2 samples and samples calcined at 350°C, 450°C and 550°C temperatures.

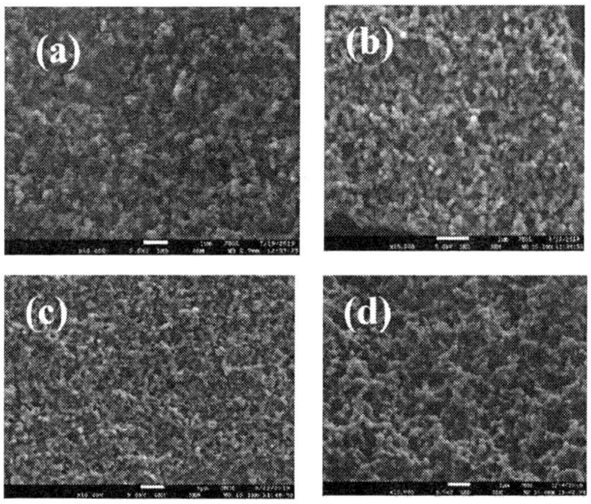

Figure 2. SEM images of a) As-Prepared b) 350oC c) 450oC d) 550oC (scalebar 1 μm)°C c) 450°C d) 550°C (scalebar 1 μm)

Corresponding author: mbuleps1@unisa.ac.za

SYNTHESIS AND CHARACTERIZATION OF REDUCED GRAPHENE OXIDE

B. Tlhaole[1,2], N.J. Coville[2,1], E.C. Linganiso[3,2,1] and J.T. Mhlongo[1,2]

[1]MSI, School of Chemistry, University of the Witwatersrand, Johannesburg, [2]DST-NRF CoE-SM, University of the Witwatersrand, Johannesburg, [3]MMU, University of the Witwatersrand, Johannesburg

Graphene-based materials and their composites have become essential in applications, such as sensors, biomedicine, energy storage devices, catalysis, waste water removal, etc.[1, 2]. Graphene is a single layer of sp^2 bonded carbon atoms in a hexagonal lattice arrangement. Reduced graphene oxide (rGO), which is part of the graphene family, can be easily obtained by reducing graphene oxide (GO). rGO is important because it has similar characteristics to pristine graphene. It has a large surface area, tuneable conductivity, is chemically stable and makes bulk synthesis of graphene possible[1-3].

Reduction of GO to rGO can be achieved by employing various methods including chemical vapor deposition, solvothermal and hydrothermal methods as well as electrochemical reduction[5]. Researchers are trying to move towards using efficient and environmentally friendly techniques that can produce similar results with regards to the physical, chemical and electronic properties as the conventional ones. Microwave-assisted (MW) synthesis techniques have gained interest over the last few decades because of the quick reaction time, simple medium, short time to reach the desired temperature and the ability to control the shape and size of nanomaterial produced[4].

In this study, graphene oxide was firstly synthesized using the Improved Hummers method[6] and then reduced to obtain reduced graphene oxide. The reduction was carried out by exposing the GO to a MW reduction strategy using hydrazine, vitamin C and hydrogen peroxide as possible reducing agents. A conventional chemical vapor deposition method was also applied for comparison. The structural morphological and surface area properties of the as-synthesized material was studied using X-ray diffraction (XRD), Raman spectroscopy, transmission electron microscopy (TEM), energy dispersive X-ray diffraction spectroscopy (EDS), X-ray photoelectric spectroscopy (XPS) and Brunauer-Emmer-Teller (BET).

XRD revealed the amorphous nature of the material with prominent (002) and (100) carbon peaks. The TEM micrograph in Figure 1 shows sheets obtained via MW reduction with hydrazine as reducing agent. The rGO appears as a single sheet with wrinkles due to the defects present in the carbon structure as revealed by Raman. EDS and XPS data showed a decrease in the amount of oxygen from GO to rGO. This suggests that the method was successful in reducing the GO. Comparable results were also obtained when the GO was exposed to only microwave irradiation with no reducing agent being employed.

Therefore, we can conclude that MW reduction has shown to be effective in reducing GO to rGO. The single layer sheets obtain using this technique are comparable to those obtained using a conventional method.

References:
1. Mohan, V.B. *et al.* (2018) Compos Part B-Eng 142, 200.
2. Lim, J.Y. *et al.* (2018) J. Ind. and Eng. Chem. 66, 29.
3. Zhou, L. *et al.* (2017) Chem. Phys. Lett. 677, 7.
4. Hasanpoor, M., Aliofkhazrei, M. and Delavari, H. (2015) Procedia Materials Science 11, 320.
5. Lavin-Lopez, M.P. *et al.* (2017) ADV POWDER TECHNOL 28(12), 3195.
6. Marcano, D.C. *et al.* (2010) ACS Nano, 4(8), 4806.

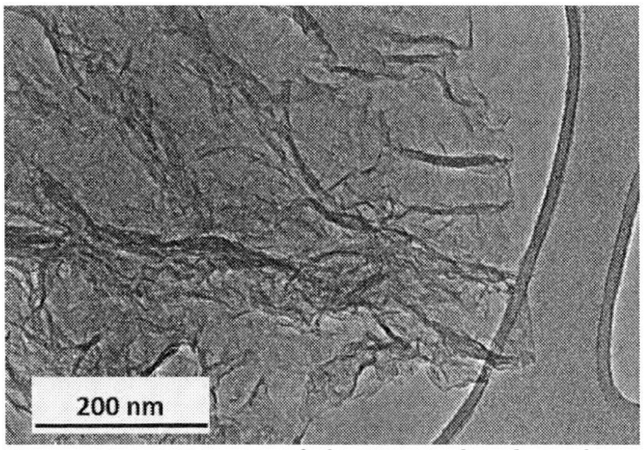

Figure 1. TEM image of the MW reduced graphene oxide using hydrazine as reducing agent.

Corresponding author: tlhaoletumi@gmail.com

THE EFFECT OF SHI-IRRADIATION ON THE MICROSTRUCTURES OF YAP, YAG & YIG.

A. Janse van Vuuren[1], V.A. Skuratov[2], G. Aralbayeva[3], A. Dualetbekova[3], M. Zdorovets[4] and J.H. O'Connell[1]

[1]Centre for HRTEM, Department of Physics, Nelson Mandela University, Port Elizabeth, South Africa, [2]Flerov Laboratory for Nuclear Research, Joint Institute for Nuclear Research, Dubna, Russia, [3]L.N. Gumilyov Eurasian National University, Astana, Kazakhstan, [4]Astana Branch of Institute of Nuclear Physics, Astana, Kazakhstan

The radiation stability of yttrium-based oxides as important ingredients of oxide dispersion strengthened alloys is a subject of extensive study. Oxide particles containing yttrium have been shown to be very stable at elevated temperatures[1]. Of particular interest to investigations in this field is the radiation damage induced by swift heavy ion (SHI), which serves to simulate fission fragment impact. The aim of this investigation is to assess the effects of SHI irradiation on the microstructure of single crystalline YAP (Y-Al-Perovskite, $YAlO_3$), YAG (Y-Al-Garnet, $Y_3Al_5O_{12}$) and YIG (Y-Fe-Garnet, $Y_3Fe_5O_{12}$).

Irradiation was conducted at the FLNR, JINR in Dubna, Russia with the IC-100 cyclotron and the DC-60 cyclotron at the IRC in Astana, Kazakhstan. TEM lamellas were prepared using an FEI Helios Nanolab FIB. Specimens were analysed with either a JEOL JEM 2100 LaB_6 or a JEOL ARM200F TEMs operated at 200 kV.

The samples used in this investigation were irradiated with Ar, Bi, Kr and Xe ions with energies ranging from 46 MeV to 2.6 GeV to fluences in the range of 10^{10} - 10^{14} cm^{-2}. A subset of these samples was irradiated at temperatures ranging from LNT to 700 °C. The overall intention of these experiments was to determine the threshold stopping power for latent track formation and the influence of increasing stopping power and irradiation temperature on track morphology, employing TEM techniques. Since these materials are similar in elemental composition but differ in crystal structure, the differences in microstructure of latent tracks is very informative when considered within the framework of the thermal spike model[2].

An example of latent ion tracks in the three materials is shown in Fig. 1 a-c. High angle annular dark field (HAADF) scanning TEM micrographs of planar samples, irradiated with 167 MeV Xe (similar stopping powers), show that latent tracks in YIG (Fig 1a) are the largest, slightly smaller in YAG (Fig. 1b) and smallest in YAP (Fig. 1c). These images fit well with the general trend, i.e. track diameters are largest in YIG and smallest in YAP at the similar stopping powers with tracks in YAG closer in size to those in YIG.

The results of this investigation revealed that amorphous latent ion tracks form YIG, YAG and YAP at stopping powers in the range from 11 to 41 keV/nm. Track diameters also increase with increasing stopping power and fit well with results from literature where available. YAP also consistently has latent ion tracks smaller in size than those in YAG and YIG at similar stopping powers. The results from samples irradiated at different temperatures suggest that approximately 10 keV/nm more energy is required to produce tracks, in YAP, that are similar in size to those in YAG and YIG.

The results also suggest that the complexity of the crystal structure plays a significant role in the recrystallisation behaviour of some crystals, this effect has been observed before in other materials[3,4].

References:
1. Ramar, A. *et al.* (2009) J. Nucl. Mater. 386, 515.
2.Toulemonde, M. *et al.* (2000) Nucl. Instr. and Meth. in Phys. Res. B 903, 166.
3. Park, S. *et al.* (2015) Acta Materialia 93, 1.
4. Rymzhanov, R.A. *et al.* (2019) Sci. Rep. 9, 3837.

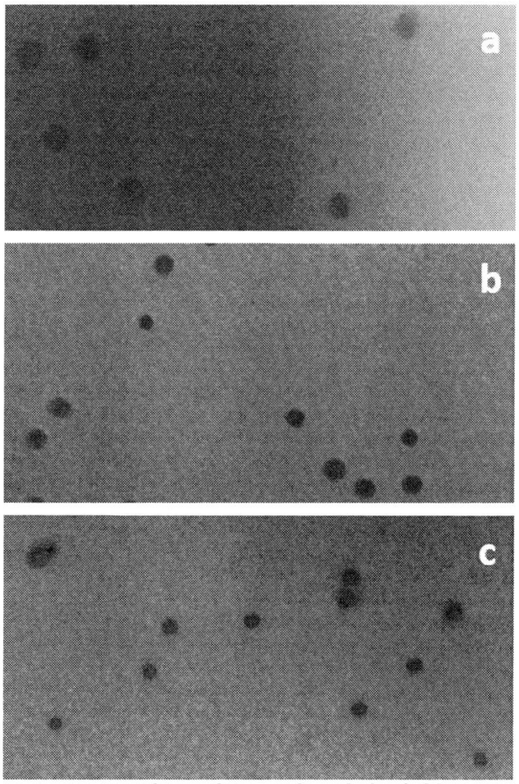

Figure 1. High angle annular dark field STEM micrographs of (a) YIG, (b) YAG and (c) YAP irradiated with 167 MeV Xe to a fluence of 5×10^{10} cm^{-2} showing latent ion tracks in a planar orientation.

Corresponding author: arnojvv@gmail.com

THE EVOLUTION OF MICROSTRUCTURES IN HIGH VANADIUM ALLOYS

A.D. Mabeba[1,2] and C.W. Siyasiya[3]

[1]Mintek, Pretoria, [2]University of Pretoria, Pretoria, [3]Materials Science and Metallurgical Engineering, University of Pretoria, Pretoria

Comminution or particle size reduction plays a major role in the extraction of minerals. Comminution includes the use of crushing and grinding equipment where abrasion wear resistance is a major requirement. The use of durable industrial equipment and raw materials reduces downtime, repair costs, and replacement costs. The introduction of a hard carbide phase is usually considered to improve abrasion wear properties. Chromium (Cr) has been widely successful for this application by forming M_7C_3 and $M_{23}C_6$ to support the abrasion wear properties of the martensitic matrix.

Vanadium (V) produces the hardest carbides (MC) and similar to Cr, V can also ensure the stability of a carbide phase. In this study the influance of carbon (C) and V addition on abrasion wear properties and microstructural evolution of VC was studied.

The alloys were melted in an induction furnace where they were cast into sand moulds after superheating to 1550°C. The alloys were solution treated at 1150°C and 1250°C, quenched in oil and tempered between 300°C and 600°C at 100°C intervals. Vanadium carbides form as VC from the reaction V + C = VC, although a large degree of non-stoichiometry has been reported between $VC_{0.74}$ and $VC_{0.91}$ [1]. For the reaction:
V + C = VC, the ratio of V to C is 4.2413:1,
And for the reaction: V + 0.74C = VC0.74 the ratio becomes 3.1412:1.
C needs to be the excess to be able to strengthen the matrix. For a ratio that does not favor excess carbon the matrix becomes ferritic. The formation of VC depletes C from the matrix. The as-cast microstructures of 1.54mass% C and 6.84mass%V (ratio 4.44:1) is shown in Figure 1 (a) after etching with 2% nital.

When the C excess is catered for, a harder matrix was able to form. The VC nucleated as spheroids and the volume fraction of carbides increases with increasing V and C additions. Increased alloy additions increase the size of the carbides by dendritic growth. The carbides increase their size by heterogeneously nucleating on existing carbides to create flowerlike structures, Figure 1 (b) (2.91mass% C and 11.0mass%V). Figure 1 (b) was etched with Le-Pera (equal mixtures of 1g $K_2S_2O_5$ in 100ml of water and $C_6H_3N_3O_7$in ethanol).

The dendritic structures have been reported to be eliminated by rare earth modifications. The rare earth metals scavenge sulphur (S) and phosphorus (P) which segregates on the corners of the solidifying VC particles to promote dendritic growth through heterogeneous nucleation[2-4]. Increased soaking temperatures spheroidise the carbides by dissolving carbide edges as they reduce their surface energy. Figure 3 is a micrograph of a 2.91mass%C and 6.08mass%V that was soaked at 1250°C and quenched in oil. The micrograph was etched with Le-Pera and high carbon martensite plates are coloured blue. The lower carbon martensite lathsare coloured brown, the austenite is unattached (white) with carbidesmostly spherical.

Hardness and abrasion wear properties improves when excess carbon is added and solution treatment temperature is increased to 1250°C.

References:
1. Shattynski S.R. (1978) J. Oxidation of Metals 13(2), 105.
2. Fleishchmann A.H. (1995) The manufacture and investigation of abrasion of abrasion resistant FE-V-C alloys.
3. Efremenko V.G. *et al.* (2013) J. *Friction and Wear*, 34(6), 466.
4. Wang C.C, Hsu H.T, Ma Q. (1990) Materials Science and Technology 6(9), 905.

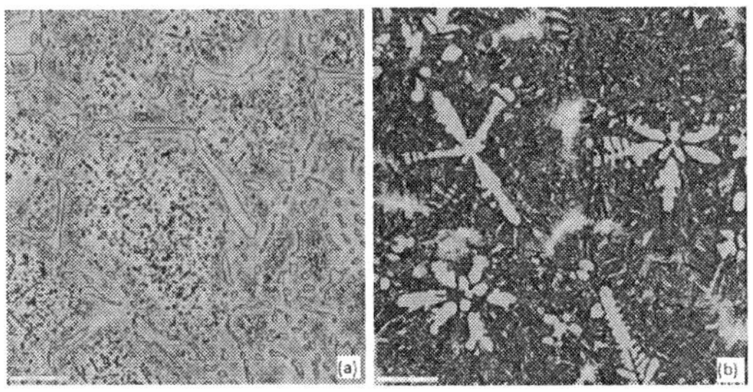

Figure 1. (a) General etching of VC and matrix using 2%nital and, (b) Brown martensitic matrix, light retained austenite and dendritic/flowerlike VC, etched with Le-Pera

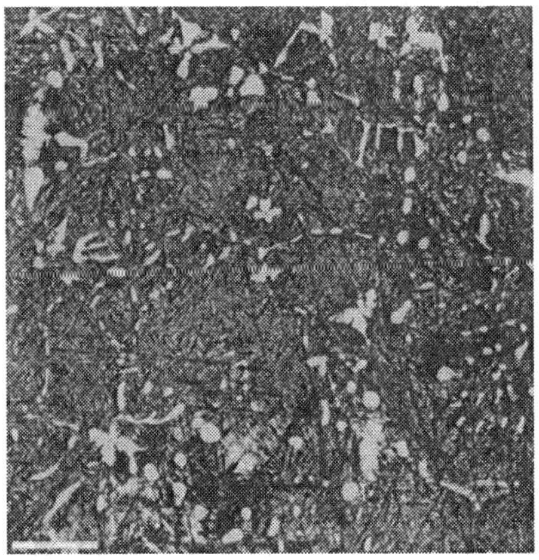

Figure 2. Brown colour martensitic matrix, blue carbon rich martensite needles, white retained austenite and VC, etched with Le-Pera.

Corresponding author: absalomm@mintek.co.za

TIN OXIDE THIN FILMS BY THERMAL OXIDATION

A.F. James[1], C.J. Arendse[1], C.J. Oliphant[2], L.F. Fourie[1], S. Ngqoloda[1] and S. Magubane[1]

[1]Dept. of Physics & Astronomy, University of the Western Cape, Cape Town, [2]Materials Characterization, NMISA, Pretoria

A vast amount of solar cell research interest is focused on organic-inorganic halide perovskite solar cells (PSCs), with its power conversion efficiency at 22%[1]. A vital component in PSCs is the electron transport layer (ETL). Tin dioxide (SnO_2) thin films are a worthy candidate for as ETL due to its suitable energy level, high electron mobility of 240 $cm^2v^{-1}s^{-1}$, desirable band gap of 3.6 – 4.0 eV, and ultimately proves to be a low temperature technique for ETL production[2]. A variety of methods are available to prepare SnO_2 thin films, such as spin coating and chemical bath deposition. However, the customary solid-state method, which incorporates thermal decomposition and oxidation of a metallic Sn precursor compound in an oxygen abundant atmosphere prevails to be low-cost, repeatable and allows for large-scale processing.

In this work, Sn_xO_y thin films were prepared by thermal oxidation of a Sn thin film, as follows: A 40 nm-thick Sn thin film was deposited on a crystalline Si (100) substrate by thermal evaporation with a deposition rate of 0.7 Å/s. To induce oxidation, the Sn thin film was subsequently annealed in ambient air for 2 hours at temperatures ranging from 200 – 400 °C. We report on the effect of annealing temperature on the evolution of the structural, vibrational and morphological properties of the Sn thin film. The crystallinity of the samples was investigated by x-ray diffraction, using a PANalytical Empyrean x-ray diffractometer. Fourier transform infrared (FTIR) was used to investigate the vibrational properties of the thin films, using a Perkin Elmer FTIR spectrometer. The morphology evolution of the surface was probed using a Zeiss Cross Beam 540 Focused Ion Beam Scanning Electron Microscope.

Figure 1(a) shows the x-ray diffractogram of the as-deposited Sn thin film and the annealed samples. The SnO tetragonal (P4/mnm) structure and the SnO_2 tetragonal (P42/mnm) structure are present in the spectra. The SnO2 peaks are more pronounced after the 300 °C anneal, confirming the successful oxidation of Sn. The extracted lattice constants correlate to theoretical values, which confirms the structural quality. Fig. 1(b) displays the FTIR spectra of the oxidized Sn thin films at 300 - 400 °C. The peaks present at 670 – 660 cm^{-1}, 530 cm^{-1}, and 440 – 420 cm^{-1} are due to the Sn-O-Sn, Sn-OH, and Sn-O vibrations, respecively, which confirms SnO and SnO_2 formation.

Figure 2(a) - (d) displays the planar SEM micrographs of the as-deposited Sn thin film and the films annealed from 200 – 400 °C. The as-deposited Sn film is composed of closely packed irregular shaped grains with well-defined grain boundaries, which act as pathways for oxygen diffusion required for the oxidation of the metallic Sn[3]. After annealing at 300 °C, a more defined morphology is present, with large Sn_xO_y particles of size 303.71 ± 77.02 nm. Furthermore, smaller Sn_xO_y spherical particles are present at the grain boundaries, where the exposed Si substrate is evident. Defects along the Sn grain boundaries are responsible for the agglomeration and oxidation, resulting in larger sized particles. After annealing at 300 °C, the thickness and surface roughness increased, which is ascribed to the unit cell volume expansion of the Sn to Sn_xO_y phases, where Sn cations and oxygen anions diffuse between the oxide/metal and oxide/gas interfaces[3].

In summary, 300 °C was identified as the optimal annealing temperature for the formation of an SnO_2 thin film, for use as the ETL in PSCs. Furthermore, the porous morphology of the thin film is sufficient to promote the integration of the perovskite layer onto the ETL that will promote enhanced charge-transfer between the layers.

References:
1. Shi, Z. and Jayatissa, A. (2018) Materials 11(5), 729.
2. Jung, H.S. and Park, N.G. (2015) Small 11, 10.
3. Zima, T. and Bataev, I. (2016). Journal of Solid State Chemistry, 243, 282.

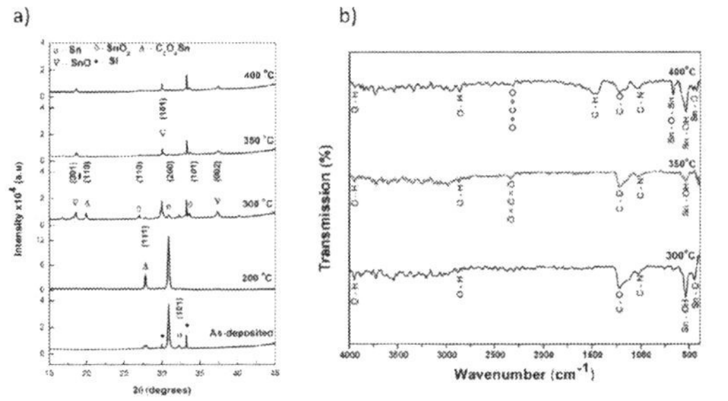

Figure 1. (a) XRD diffractograms of the as-deposited Sn thin film and the annealed samples. (b) FTIR spectra of the annealed samples from 300 – 400 °C.

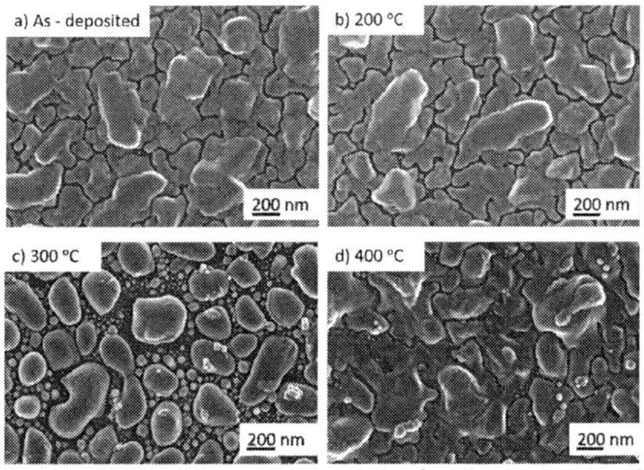

Figure 2. SEM planar micrographs for the as-deposited and annealed thin films.

Corresponding author: 3335164@myuwc.ac.za

TIN-LEAD IODIDE FORMATION BY POST-DEPOSITION ANNEALING

S.S. Magubane[1], C.J. Arendse[1], S.C. Klue[1], C.J. Oliphant[1,2] and A.S. Bolokang[1,3]

[1]Dept. of Physics and Astronomy, UWC, Bellville, [2]Materials Characterization, NMISA, Pretoria, [3]Material Science and Manufacturing, CSIR, Pretoria

In less than a decade, the power conversion efficiency (PCE) of perovskite solar cells (PSCs) has tremendously increased from 3.8 % to 23.7 %[1]. Lead-based perovskite has been at the center of this development due to its novel optoelectronic properties such as strong absorption coefficient (~10^5 cm^{-1}), the low exciton binding energy (~20 meV), and relatively long carrier diffusion length (>1 μm) [2]. However, these devices have not been commercialized yet due to issues of stability and toxicity of lead-based perovskite[2]. As a result, researchers have embarked on a search for a stable and less toxic perovskite structure. Unfortunately, stable and/or less toxic perovskite material such as tin (Sn), germanium (Ge), antimony (Sb), bismuth (Bi) etc. exhibit low PCEs. In an attempt to improve the stability and reduce the lead content in the devices, mixed perovskites have been considered, with the highest certified efficiency exceeding 19% obtained by Sn-Pb hybrid PSC[3]. In the current work, we are establishing the processing conditions of forming Sn-PbI_2 alloy, which will later be converted into perovskite to examine and control the effect of Sn defect on Pb-based perovskite.

Sn-PbI_2 alloy films were prepared on glass substrates, which were cleaned in hellmanex solution and isopropanol and subsequently rinsed in deionized water in an ultrasonic bath. The two-step sequential deposition was used in this study whereby a 100 nm PbI_2 film was deposited first via chemical vapour deposition (CVD), followed by the deposition of a 40 nm Sn film via thermal evaporation using the furnace tube and cluster apparatus for device application research (CADAR), respectively. The samples were then annealed in a nitrogen environment at 250 °C, 300 °C and 350 °C; each 30 minutes, to form Sn-Pb alloys.

The morphology and energy-dispersive spectroscopy (EDS) measurements of the samples were performed using a Zeiss Auriga high resolution scanning electron microscope (HRSEM). The crystal structure of the samples was examined using a D8-Advance from Bruker AXS x-ray diffractometer operated from 10° to 80° (2θ), using Cu-Kα radiation (1.5406 Å).

Figure 1 (a-c) depicts the SEM micrographs of Sn, PbI_2 and Sn-PbI_2 alloy thin films. It is observed from Fig. 1 (a) and (b) that the metallic Sn film is composed of irregular shaped nanoparticles, while the PbI_2 film forms large and smooth platelets with grain boundaries. Figure 1 (c) is the micrograph of Sn-PbI_2 alloy, which was formed by annealing the Sn/PbI_2 film at 250 °C for 30 minutes in a nitrogen environment. A clear change in morphology is observed after annealing, which signals the intermixing of the layers. This alloy has been identified as optimal for a void- and crack-free surface with a homogeneous elemental distribution throughout the film thickness. Figure 1 (d) shows the XRD patterns of the films. In addition to PbI_2 peaks located at 2θ values of 12.71°, 25.55°, 38.72° and 52.46° assigned to (003), (006), (009) and (0012) lattice planes that correspond to d-spacings of 6.94 Å, 3.49 Å, 2.32 Å and 1.74 Å, respectively, two Sn peaks located at 30.60° and 63.70° assigned to (200) and (400) lattice planes with d-spacings of 2.92 Å and 1.46 Å, respectively, are observed. Upon annealing, the metallic Sn peaks disappear and a $SnPbO_4$ peak at 26.94° assigned to the (121) diffracting planes with a d-spacing of 3.31 Å is observed. Based on the intensities of PbI_2 peaks, annealing at 250 °C for 30 minutes yields the desired alloy. EDS analysis confirms the expected elemental composition of the thin films. At temperatures above 350 °C, the XRD data exhibited an amorphous structure (not shown), which signals the evaporation of the PbI_2.

References:
1. Chan, S.H., Chang, Y.H. and Wu, M.C. (2019) Front. Mater. 6(1), 1.
2. Green, M.A., Ho-Baillie, A. and Snaith, H.J. (2014) Nat. Photonics 8(7), 506.
3. Li, C. *et al.* (2019) Adv. Energy Mater. 9(3), 1.

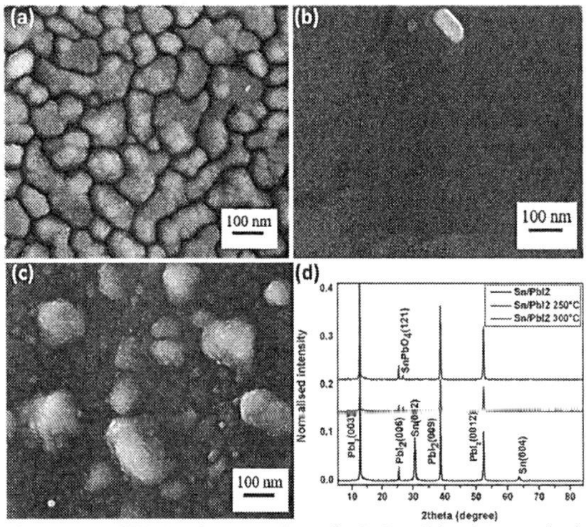

Figure 1. SEM micrographs of a) Sn, b) PbI_2 and c) Sn-PbI_2 films annealed at 250°C. d) XRD patterns of the films.

Corresponding author: 3578636@myuwc.ac.za

TiO_2 NANOWIRE ARRAYS EXPOSED TO NITROGEN PLASMA FOR APPLICATION AS ELECTRON TRANSPORT LAYERS IN PHOTOVOLTAICS

I. Cupido[1,2] and F. Cummings[1,2]

[1]Dept. of Physics, University of the Western Cape, [2]Electron Microscope Unit, University of the Western Cape

TiO_2 nanoparticle films are extremely popular electron transport layers in photovoltaics such as dye-sensitised solar cells[1], perovskite[2] and polymer hetero-junction solar cells[3]. These films, however, have significant electron trap-sties as a result of the large density of oxygen vacancies present in nano-sized TiO_2. In this regard, single crystalline TiO_2 nanowires (TiO_2 NWs) have notably fewer trapping sites that cause electron-hole recombination and photocurrent loss when used in photovoltaics. To shorten the electron transfer, vertically grown TiO_2 NWs provide the needed bridge from the top of the nanowires to the substrate during charge transport within the device.[4, 5] However, issues during the manufacturing of solar cells incorporating these structures include high temperature-dependent techniques[6] and exposure to environments with extreme levels of pH[7], all of which have a high likelihood of damaging and in some cases destroying substrates, thereby resulting in cells with poor quality and performance.

In this study, a low temperature, repeatable hydrothermal approach is taken to synthesize an array of TiO2 NWs with controlled dimensions of length and diameter. Post-synthesis, the substrates are exposed to a radio-frequency (RF) generated N_2 plasma, which allows for controlled N doping of the TiO_2 lattice; this leads to a reduction in oxygen vacancy density and electron transport promotion through the NW array. The morphology, crystalline and electronic properties of the as-synthesised NWs versus that of the N_2 plasma exposed structures are presented in this contribution.

Figure 1 (a) and (b) show plan and side-view scanning electron microscope (SEM) images of an untreated TiO_2 NW sample, in which it is shown that the general diameter and length of the wires are in the range of 68nm and 1370 nm, respectively. The SEM micrographs were collected using a Zeiss Auriga field-emission gun SEM operated at an accelerating voltage of 5 kV using an in-lens secondary electron detector. Also shown in Figure 1 are the x-ray photo-electron spectra (XPS) comparing the Ti 2p electronic structure of the untreated TiO_2 NW arrays, Figure 1 (c), to that of an array exposed to a N_2 plasma at 500W for 10 mins with the substrate at room temperature (Figure 1 (d)). The XPS results show nitrogen doping of the TiO_2 lattice, manifested by the presence of the Ti-N bond at a binding energy of 456 eV as shown by the arrow in Figure 1 (d); this is in agreement with previously reported results[8]. Doping of this nature can result in a new energy level forming in the TiO_2 bandgap. Further work to be done includes UV-Visible spectroscopy measurements to determine change in optical properties.

References:
1. O'Regan, B. and Grätzel, M. (1991) Nature, 353, 737.
2. Chen, P. et al. (2015) Journal of Energy Chemistry, 24(6), 717.
3. Ranjitha, A. et al. (2019) Materials Science for Energy Technologies.
4. M. Law, L. E. et al. (2005) Nat Mater 4 (6), 455
5. W. I. Park and G. C. Yi, (2004) Adv Mater 16 (1), 87.
6. Yu, C., and Park, J. (2010) Journal of Solid State Chemistry, 183(10), 2268.
7. Su, J., & Guo, L. (2015) RSC Advances, 5(65), 53012.
8. Chen, C.et al. (2007) J. of Nanoparticle Res. 9:365.

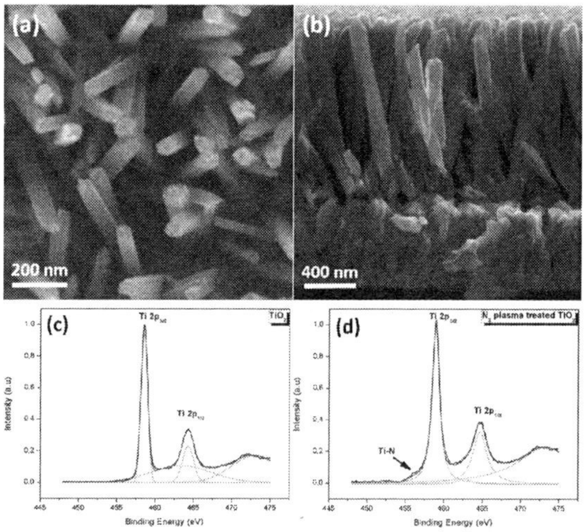

Figure 1. (a) Plan-view and (b) cross-sectional view SEM micrographs of an as-synthesised TiO2 nanowire array, grown in a hydrothermal solution for 3 hours at 150 °C; XPS of Ti 2p of (a) as-synthesised TiO2 nanowire array and (d) nanowires treated at in a N2 plasma for 10 mins at a source power of 500 W with the sample kept at room temperature.

Corresponding author: fcummings@uwc.ac.za

TOWARDS IMAGING CARBON ATOMS IN PLATINUM CARBIDES

M.J. Witcomb[1] and U. Dahmen[2]

[1]DST-NRF Centre of Excellence in Strong Materials, University of the Witwatersrand, Johannesburg, South Africa,
[2]NCEM, Molecular Foundry, LBNL, Berkeley, California, USA

Following a rapid quench from high temperature, platinum tends to form faulted dislocation loops on {100} planes instead of {111} planes as found in other fcc metals[1]. The unusual stability of these {100} defects led to the hypothesis that they were stabilized by carbon impurities[2]. A systematic study using materials of different purity confirmed this hypothesis and showed that subsequent annealing led to a series of different metastable carbide phases[3]. These phases were identified using diffraction contrast imaging, microbeam diffraction and energy loss spectroscopy. However, direct imaging of the carbon atoms by high-resolution electron microscopy has been impeded by the large mass difference between Pt and C and the sensitivity of the carbide phases to radiation damage.

This contribution presents high resolution images showing direct evidence for the presence of carbon in a bilayer carbide and explains the surprising observation that the material becomes more sensitive to radiation damage at lower accelerating voltages.

Figure 1 shows a comparison of experimental (a,b) and simulated (c,d) high resolution images taken at 800 kV. The image simulations are based on the Pt_2C structure shown in (e), in which C atoms are shown in black. In this projection, C atoms form a zig-zag arrangement with a lateral repeat distance twice that of Pt atoms. This feature is reflected in both, the experimental images (top row) and corresponding simulations (bottom row) for two different defocus settings of the microscope. The doubling of the repeat period disappears when C is removed from the simulations, thus providing direct evidence for the presence of C.

An attempt to image the C atoms directly using an aberration-corrected microscope at 80 kV was unsuccessful due to rapid radiation damage that destroyed the precipitate over the course of a single focal series. This surprising result can be explained by the enhanced energy transfer to light atoms at lower accelerating voltages. A similar enhancement of radiation damage was previously found for Li in Al alloys[4]. Calculation of the displacement scattering cross sections show that C is up to two times more susceptible to displacement damage at 80 kV compared to 800 kV, depending on its bond strength in the carbide (see Fig. 2). To determine the bond strength, equilibrium structure and stability of the different carbides of Pt will require first-principles calculations[5].

While high resolution electron microscopy at 800 kV offers direct evidence for the Pt_2C carbide phase reported in our earlier work[3], the direct imaging of C atoms requires an aberration-corrected instrument. Due to the unusual sensitivity to radiation damage, samples should either be imaged at high accelerating voltage, or preferably below the damage threshold, which for C is around 40 kV. However, to be electron transparent at such a low voltage, the sample must be extremely thin, posing a significant challenge for sample preparation.

References:
1. Westmacott, K.H. (1973) J. Phys. F: Met. Phys. 3, L217.
2. Westmacott, K.H. and Perez, M.I. (1979) J. Nucl. Mater. 83, 231.
3. Witcomb, M.J. and Dahmen, U. (2003) Proc. 2nd Int. Conf. Afr. Mater. Res. Soc. 32.
4. Rossell, M.D. *et al*. (2009) Phys. Rev. B. 80, 024110.
5. Curtarolo, S. *et al*. (2003) Phys. Rev. Lett. (2003) 91, 135503.

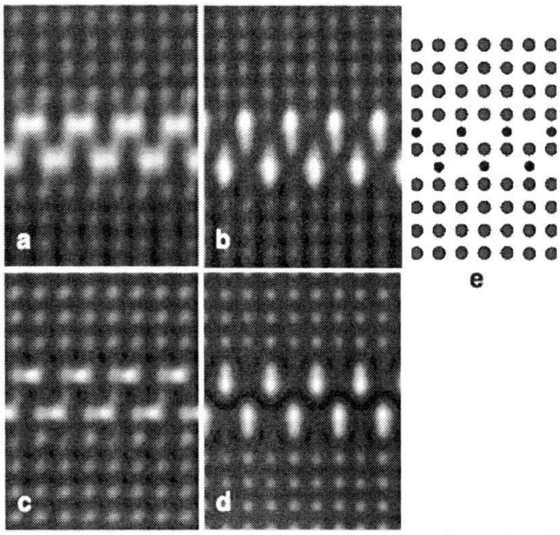

Figure 1. Comparison of observed (a,b) and simulated (c,d) images of carbide structure shown in (e).

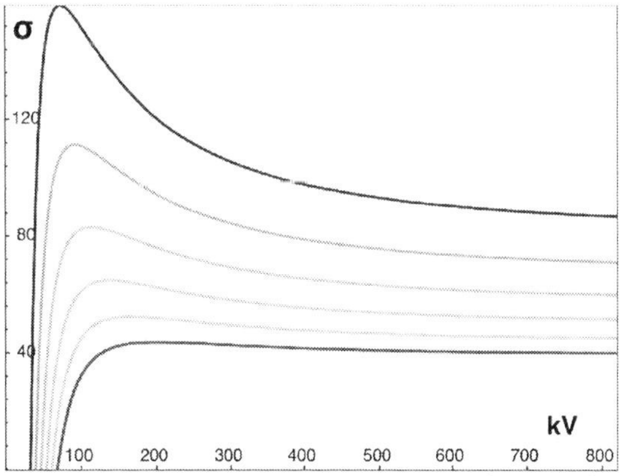

Figure 2. Scattering cross section (barns) for displacement of C atoms as a function of accelerating voltage. Different curves are for different bond strengths from 5-10 eV (top to bottom curves).

Corresponding author: michael.witcomb@wits.ac.za

ABERRATION-CORRECTED SCANNING TRANSMISSION ELECTRON MICROSCOPY OF SUPERSTRUCTURES IN NATURAL COMMENSURATE AND INCOMMENSURATE PYRRHOTITES

L. Jin[1], D. Koulialias[2,3], M. Schnedler[4], A.U. Gehring[2], M. Pósfai[5], Ph. Ebert[4], M. Charilaou[3], R.E. Schäublin[3], C.L. Jia[1], J.F. Löffler[3] and R.E. Dunin-Borkowski[1,3,4]

[1]Ernst Ruska-Centre for Microscopy and Spectroscopy with Electrons, Forschungszentrum Jülic, Jülich, Germany, [2]Institute of Geophysics, ETH Zurich, Zurich, Switzerland, [3]Laboratory of Metal Physics and Technology, Department of Materials, ETH Zurich, Zurich, Switzerland, [4]Institute for Microstructure Research, Peter Grünberg Institute, Forschungszentrum Jülich, Jülich, Germany, [5]Department of Earth and Environmental Sciences, University of Pannonia, Veszprém, Hungary

Pyrrhotites form an extended class of minerals, which have the chemical formula $Fe_{1-\delta}S$ ($0 < \delta \leq 1/8$), contain layered arrangements of ordered Fe vacancies and are important for understanding the magnetic signals that originate from certain naturally occurring rocks. Despite extensive studies over the past century, the local structural details of pyrrhotite superstructures formed by different arrangements of Fe vacancies remain unclear, in particular on the atomic scale.

The primary difficulty in characterizing pyrrhotite superstructures results from the intrinsic complexity of the Fe vacancy ordering, in particular in incommensurate polymorphs. Relevant structural details in pyrrhotite include not only the positions of Fe vacancies within the FeS framework, but also the Fe atom shifts that are associated with the presence of these vacancies. For example, Fe atoms in 4C pyrrhotite are not expected to line up coincidently when they are viewed along major zone axis directions perpendicular to the cm axis. However, direct local experimental observations are still lacking.

Accordingly, we have used atomic-resolution high-angle annular dark-field (HAADF) imaging and nanobeam electron diffraction in the scanning transmission electron microscope (STEM) to study natural pyrrhotite samples that contain commensurate 4C and incommensurate 4.91±0.02C polymorphs.

Simulated HAADF STEM images of 4C pyrrhotite viewed along $[010]_m$ and $[110]_m$-type axes are shown in Figs 1a-d for sample thicknesses of 11 nm (Figs 1a, c) and 88 nm (Figs 1b, d). The simulations show that the visualization of Fe vacancy ordering depends strongly on sample thickness (Figs 1a-d). Figure 1e shows mean intensities I measured from the simulated images for filled atomic columns of $Fe_{(F)}$ (black: $[010]_m$ and red: $[110]_m$-type projections), vacancy-containing $Fe_{(V)}$ (blue) and S (yellow) plotted as a function of sample thickness. For the $Fe_{(V)}$ and S columns, the values are essentially the same for both projections. The inset to Fig. 1e shows the ratio $I_{Fe(V)}/I_{Fe(F)}$ for sample thicknesses of up to ~90 nm. This ratio is only slightly higher than the nominal value of 0.5 used in the simulations (as a result of electron dechannelling) when the sample is thinner than ~10 nm. For higher sample thicknesses, this ratio displays a monotonic but non-linear increase in both projection directions, reaching a value of ~0.9 at a specimen thickness of ~90 nm, resulting in an intrinsic difficulty in resolving vacancy ordering in thicker samples. We performed comparisons of experimental HAADF STEM images with simulations such as those shown in Fig. 1 to interpret measurements of both the intensities and the picometer-scale shifts of individual Fe atomic columns in 4C pyrrhotite. In 4.91±0.02C pyrrhotite, we find that 5C-like nano-regions join at anti-phase-like boundaries. Our results provide atomic-scale evidence for the presence of nano-regions with different displacive sequences of Fe column shifts in incommensurate 4.91C pyrrhotite and reveal that structural (i.e., displacive) anti-phase-like stackings between unequally-sized nano-regions are the primary cause of incommensurability.

References:
Jin, L. *et al.* In preparation for submission to Am. Mineral.

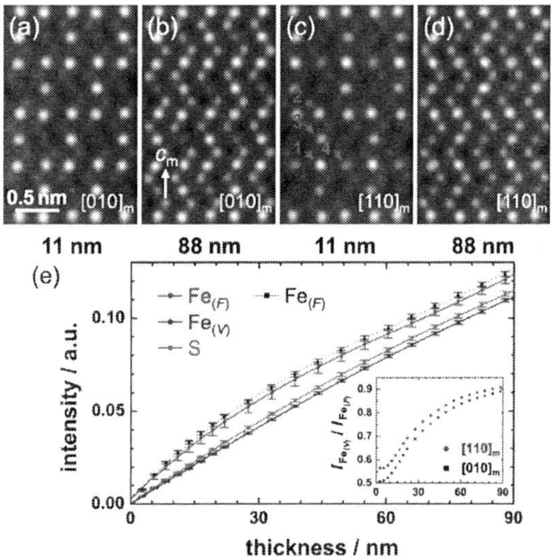

Figure 1. (a-d) Simulated HAADF STEM images viewed along (a, b) $[010]_m$ and (c,d) $[110]_m$-type directions for sample thicknesses of 11 and 88 nm. The numbers 1-4 mark four types of filled Fe(F) column that have increasing atomic splittings in projection. (e) Mean intensities I in the simulated images at the positions of individual atomic columns of $Fe_{(F)}$ (black: $[010]_m$; red: $[110]_m$), $Fe_{(V)}$ (blue) and S (yellow) plotted as a function of sample thickness. The inset shows the ratio between the intensities of the $Fe_{(V)}$ and $Fe_{(F)}$ columns plotted as a function of sample thickness. A ratio close to 0.5 is maintained for sample thicknesses of up to ~10 nm.

Corresponding author: r.dunin-borkowski@fz-juelich.de

COMPUTATIONAL MATERIALS MODELING: FROM PROCESSING TO PROPERTIES

M.R. Rolchigo[1], M.A. Orondo[2,3] and R. LeSar[2,3]

[1]Lawrence Livermore National Laboratory, [2]Materials Science and Engineering, [3]Ames Laboratory, Iowa State University, USA

Computational modeling has advanced sufficiently to have become an integral part of materials research, with modeling tools available to describe phenomena at almost all relevant length and time scales. The focus of our work is to use such modeling to explore the so-called "materials tetrahedron", which is used to visually represent the close connection between processing, structure, properties, and performance in a material. Our work is concentrated on the processing-structure-properties part of the tetrahedron, modeling how changes in process conditions affect the behavior of the as-processed material. More specifically, we use modeling of the additive manufacturing (AM) processing to predict as-processed microstructures, which serve as input for other models that we use to predict mechanical behavior.

We use a coupled cellular automata approach to examine the role of both processing conditions and alloy composition on as-processed AM microstructures. Heat transport, melting, and fluid flow are modeled at the macroscale via a thermal Lattice Boltzmann model (or other continuum-level models), accounting for laser absorption and Marangoni flow in the melt pool[1]. At the microscale, solidification is modeled using a cellular automata model for the solidification, with temperature and fluid flow boundary conditions passed from the macroscale, enabling the linking of process parameters, melt pool conditions, alloy thermodynamics, and microstructure[1]. We used these models to predict solidification morphology as a function of the melt pool conditions for a series of binary (and some ternary) alloys. In Fig. 1 we show an example that illustrates the large impact that alloy system can have on as-processed microstructures. We have studied both titanium and steel alloys with this approach.

We have recently developed a simulation method that directly couples polycrystal plasticity (PP) calculations with discrete dislocation dynamics (DDD) simulations for modeling the deformation of microstructures with arbitrary grain size, shape, and orientation[2-4]. Based on an eigenstrain formalism and solved with fast Fourier transforms (FFT), the FFT-PP-DDD approach is sufficiently fast computationally to enable detailed studies of the role of dislocation substructure development in the deformation of polycrystalline materials. In Fig. 2 we show an example of a calculated dislocation microstructure in a system of randomly oriented grains. We thus can predict the effects of dislocation substructure development on the stress-strain behavior in polycrystalline materials. We are also using this approach to study the high-temperature failure in steel alloys for energy applications. We will show how we can use the FFT-DDD method to extract the formation energy of voids in dislocated systems, from which void nucleation and growth models can be constructed. We will use the results of these simulations to develop a model for ductile failure.

We thus have the basic tools needed to reach our goal of creating a direct link between process conditions and mechanical behavior. Our plan is to use these tools to help guide the development of alloy systems for the additive manufacturing of materials for selected applications. We are developing an approach in which the results of the modeling will be directly linked with experiment using modern machine learning tools. Though these efforts, we will close the gaps in our models and decrease the uncertainty in our predictions.

References:
1. Rolchigo, M. R. and LeSar, R. (2019) Comput. Mater. Sci. 163, 148.
2. LeSar, R. and Capolungo, L., Handbook of Materials Modeling, Andreoni, W. and Yip, S. (editors) (Springer, Cham, 2018).
3. Bertin, N. *et al.* (2015) Modelling Simul. Mater. Sci. Eng. 23(6), 085005.
4. Graham, J., Rollett, A. D. and LeSar, R. (2016) Modelling Simul. Mater. Sci. Eng. 24(8), 085005.

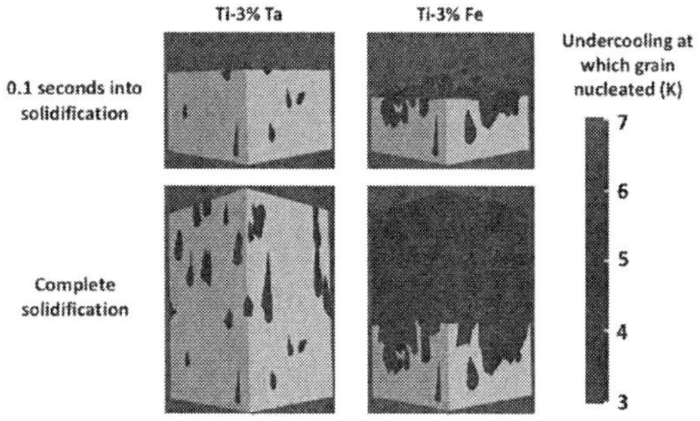

Figure 1. Predicted microstructure of two titanium alloys: Ti-3% Ta on the left and Ti-3% Fe on the right under the same processing conditions.

Figure 2. Calculated dislocation substructure in a polycrystalline material with randomly oriented grains. Dislocations are colored by their grains.

Corresponding author: lesar@iastate.edu

COUPLED TEM AND ATOM PROBE INVESTIGATION OF PHASE INVERSION IN A HIGH ENTROPY ALLOY

R. Banerjee[1], V. Soni[1], S. Dasari[1], O.N. Senkov[2], Y. Zheng[3], D.B. Miracle[2] and H.L. Fraser[3]

[1]Materials Science and Engineering, University of North Texas, Denton, USA, [2]Air Force Research Laboratory, Ohio, USA, [3]CAMM, The Ohio State University, Columbus, USA

This paper will discuss, presumably, the first experimental report on the phenomenon of "phase inversion" in metallic alloys. Such a phase inversion phenomenon has been experimentally observed in the case of a refractory high entropy alloy (RHEA), $Al_{0.5}NbTa_{0.8}Ti_{1.5}V_{0.2}Zr$[1]. Phase inversion, usually observed in polymer blends, is driven by difference in viscosity of the two phases driven by the tendency of the higher viscosity phase compact itself to minimize the elastic energy of the system. Phase inversion in crystalline solid systems is driven by the differences in elastic modulus of the two phases. Quenching from a high temperature single phase field, the RHEA exhibits a co-continuous mixture of a disordered BCC and an ordered B2 phase, as shown in Fig. 1, reminiscent of the early stages of a spinodally decomposed microstructure. The relatively lower magnification aberration-corrected HAADF-STEM image (Fig. 1(a)) shows the interconnected nature of the two phases, while the higher magnification image in Fig. 1(b) highlights a specific B2/BCC interface. Upon isothermal annealing at 600 °C this co-continuous mixture of B2 and BCC develops further via spinodal-like decomposition into a continuous B2 matrix with discrete cuboidal BCC precipitates aligned along the <001> directions. Longer term annealing at 600 °C results in the development of necking constrictions along the B2 channels, eventually pinching-off these channels and making the BCC phase continuous with discrete B2 precipitates. This inversion process can be related to the simultaneous operation of two processes: (i) spheroidization of the initially discrete cuboidal BCC precipitates driven by a reduction in the total interface energy and (ii) an increase in the stiffness of the B2 phase, relative to the BCC phase, due to chemical composition changes during annealing, forcing the B2 regions to become discrete driven by the reduction in the total elastic strain energy.

Different stages of spinodal decomposition and subsequent coarsening leading to phase inversion were characterized by coupling transmission electron microscopy (TEM) and atom probe tomography (APT) techniques. APT reconstructions from different intervals during the isothermal annealing, revealed a continuous change in the composition of the two phases present, confirming a spinodal decomposition. The composition of the phases reaches near equilibrium at ~5 hrs of annealing at 600 °C and remains constant thereafter. During coarsening, the formation of bulges in the B2 phase at the intersection of continuous B2 channels, leads to thinning of the channels at central regions. Eventually the two BCC cuboids adjacent to a B2 channel coalesce, pinching off the B2 channel at the center, making the BCC phase continuous. The phenomenon of phase inversion in this alloy significantly change the mechanical response under compression, leading to a better balance of strength and ductility[2].

References:
1. Soni, V. *et al.* (2019) Acta Mater. (under review).
2. Soni, V. *et al.* (2018) Sci. Rep. 8(1), 8816.

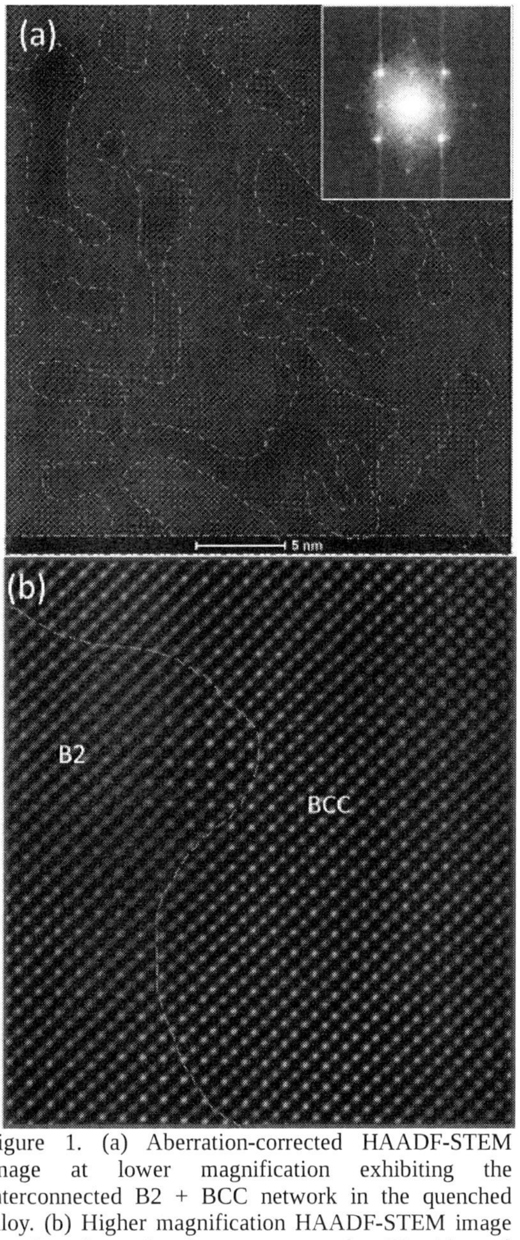

Figure 1. (a) Aberration-corrected HAADF-STEM image at lower magnification exhibiting the interconnected B2 + BCC network in the quenched alloy. (b) Higher magnification HAADF-STEM image showing the ordered structure on the B2 side and disordered structure on the BCC side.

Corresponding author: raj.banerjee@unt.edu

DEVELOPMENTS IN HIGH SPEED STRUCTURAL IMAGING OF LOW DIMENSIONAL MATERIALS

A.I. Kirkland[1,2,3]

[1]Dept. of Materials, University of Oxford, [2]ePSIC, Diamond Lightsource, [3]The Rosalind Franklin Institute, Oxford, UK

This lecture will describe firstly recent developments in the use of Scanning Transmission and Transmission Electron Microscopy applied to studies of dynamics in low dimensional materials. I will describe recent work using high speed direct electron detectors and artificial intelligence / machine learning to automatically map defect and adatom migrations in graphene from large data sets. I will then show how this approach can be extended to probe the local kinetics of defect transitions. Finally, I will discuss the use of similar detectors in electron ptychography, in particular under extremely low dose conditions using binary counting.

The development and commercialization of direct electron detectors suitable for use at intermediate electron energies as typically employed in the transmission electron microscope has led to significant progress in imaging, diffraction and spectroscopy[1]. In the life sciences improved detection efficiency has made electron microscopy the most rapidly growing structural method used in studies of biological structures[2]. In the physical sciences improved detector resolution due to a lack of scattering in a coupled scintillator and increased frame rates have enabled high resolution imaging of dynamic processes and have supported new experimental geometries including 4D STEM[3].

To usefully deploy Graphene and related materials in electronic applications[4-6] it is essential to understand the behavior of defects, which have been the subject of extensive research in silicon devices for decades. Moreover, instrumental advances make it possible to image these defects at primary energies below those that cause significant specimen damage whilst retaining sufficient spatial resolution to resolve the local atomic configurations around the defect site[7]. Combining these electron-optical developments with new high-speed detectors has further advanced this field and defect motions and rearrangements can now be tracked with ms timings.

However, the extremely large datasets (typically 106 images or greater) that can be routinely acquired makes conventional manual image processing, with human intervention intractable. Theses also put a hard restriction on the scalability of data processing in some important areas, in particular where the collection of large image series is routine. I will describe how this can be overcome using a deep learning neural network[8] which performs atomic model abstraction from low dose high framerate experimental graphene images (Fig. 1). Although the training of such neural networks requires significantly more effort than classical image processing, this method is more general can autonomously process large datasets and can readily be extended to studies of other two-dimensional materials. Using this approach, it is possible to identify many instances of specific defect transitions and to map the lifetimes of defect states. In turn these can be used as input to density functional theory to model the potential energy landscape for the transitions.

Finally, I will highlight the use of fast detectors for electron Ptychography at low dose. Acquisition of a ptychographic dataset requires the collection of a series of far field diffraction patters as a function of probe position at the specimen plane. This dataset can then be used to recover the complex specimen object function using either an iterative or non-iterative algorithm. Importantly, ptychography is a dose efficient technique, enabling effective phase reconstruction of radiation sensitive samples. At low dose the sampling of the diffraction pattern in the far field is sparse and a counting direct electron detector can be operated in a binary mode to provide an effective speed increase. I will illustrate this method using examples of radiation sensitive mesoporous materials.

References:
1. Mir, J.A. *et al.* (2017) Ultramicroscopy 182, 44.
2. Callaway, E. (2015) Nature 525, 172.
3. Gao, S. *et al.* (2017) Nature Comm. 8, 163.
4. Novoselov, K. S. *et al.* (2012) Nature 490, 192.
5. Schwierz, F. (2010) Nature Nanotech. 5, 487.
6. Britnell, L. *et al.* (2012) Science 335, 947.
7. Robertson, A.W. *et al.* (2012) Nature Comm. 3, 1144.
8. LeCun, Y. *et. al.* (2015) Nature 521, 436.

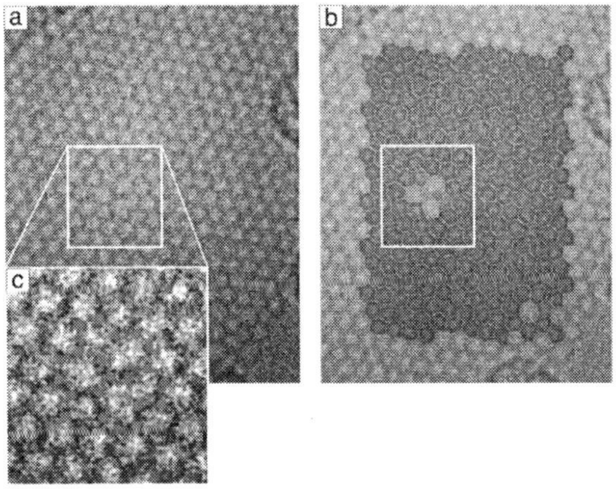

Figure 1. (a) Low-dose image of a graphene sheet recorded with a 1ms exposure at 80 kV containing defects; (b) Automatically generated annotation with 5-membered rings (pink), 6-membered rings (green), and 7-membered rings (yellow) overlaid on top of the experimental image; (c) enlarged defect area with carbon atom positions marked by blue circles.

Corresponding author:
angus.kirkland@materials.ox.ac.uk

INTEGRATED DIFFERENTIAL PHASE CONTRAST (iDPC) STEM FOR LOW Z DETECTION AND FOR LOW-DOSE IMAGING APPLICATIONS

A. Yalcin

Thermo Fisher Scientific, Eindhoven, The Netherlands

One of the main challenges in conventional STEM techniques is the difficulty to fully image and interpret a lattice consisting of both high and low Z elements. While only high Z elements can be imaged in HAADF-STEM, (A)BF-STEM can display low Z elements, though atomic positions in the images are difficult to interpret as there is no clear contrast variation between different atoms[1]. Novel integrated differential phase contrast (iDPC) STEM imaging is instrumental in showing both low Z and high Z elements with clear contrast variation[2].

iDPC STEM is based on center of mass (COM) STEM imaging[3], in which the electric field projected on the electron beam by the specimen leads to the COM displacement of CBED disc (Fig. 1). This displacement can be depicted as a 2D vector on the detector plane and the x- and y- components of this vector can be detected with a segmented STEM detector. These two images correspond to conventional DPC STEM images (Fig. 2). The electric field is conservative vector field and is the gradient of the electrostatic potential field of the sample, therefore the latter can be retrieved using the integration step (Fig. 2).

The strength of iDPC STEM in simultaneous high and low Z detection has been shown experimentally in GaN[2,4] and Ɣ-TiH5 domains. Moreover, iDPC STEM images have also been shown to exhibit improved contrast in beam sensitive materials, exemplified with graphene[2].

iDPC STEM imaging is further strengthened on the recently launched Thermo Scientific Spectra high end S/TEM platform with the S-CORR 5th order probe corrector (patent US20110114852A1) and with the new Panther STEM detection system. S-CORR corrects six-fold astigmatism at all high tensions whereby smaller probe sizes and better STEM resolution can be achieved. Furthermore, with the Panther STEM detection system delivering up to 100x more gain, high contrast iDPC STEM imaging can now be achieved even in life sciences dose range (<50 $e^-/Å^2$) allowing imaging very beam sensitive specimens such as zeolites and MOFs.

In the presentation, several application examples will be shown, clearly displaying the strength of iDPC STEM in combination with S-CORR probe corrector and Panther STEM detection system in low Z detection and in high contrast low dose imaging applications.

References:
1. Bosch, E.G.T. and Lazić, I. (2015) Ultramicroscopy 156, 59.
2. Lazić, I., Bosch, E.G.T. and Lazar, S. (2016) Ultramicroscopy 160, 265.
3. Müller, K. *et al.* (2014) Nat. Commun. 5, 5653.
4. Yücelen, E., Lazić, I. and Bosch, E.G.T. (2018) Sci. Rep. 8, 2676.
5. de Graaf, S. *et al.* arXiv:1812.09118.

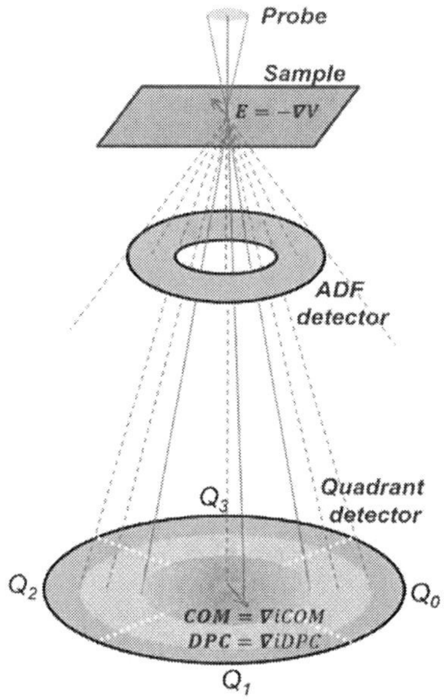

Figure 1. iDPC STEM image formation. Center of Mass (COM) movement of the CBED pattern at the detector plane due to the local electric field of the sample.

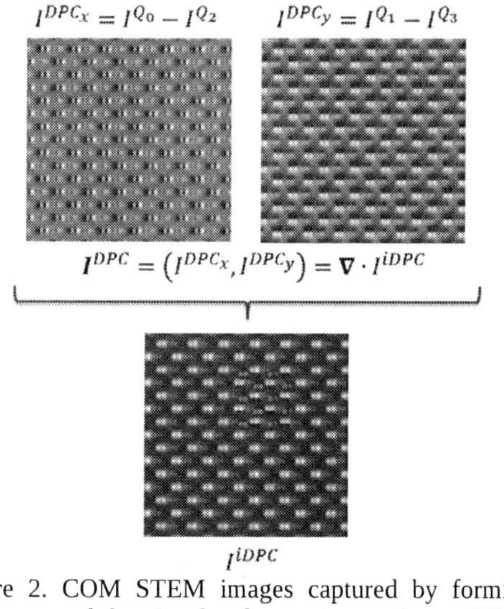

Figure 2. COM STEM images captured by forming differences of the signals of opposite quadrants (DPC) and an integration step (iDPC STEM). Example images are of GaN [211].

Corresponding author: anil.yalcin@thermofisher.com

MICROSTRUCTURE AND FATIGUE CRACK GROWTH MECHANISMS OF TI-6AL-4V FABRICATED BY LASER POWDER BED FUSION

T.H. Becker[1], N.M. Dhansay[1] and K. Vanmeensel[2]

[1] Department of Mechanical and Mechatronic Engineering, Stellenbosch University, Stellenbosch, [2]Department of Materials Engineering, KU Leuven, Belgium

Over the past decade, additive manufacturing (AM) of Ti-6Al-4V has gained significant attention in academia and industry alike. This two-phased alloy permit structure refinement through, for example, heat-treatment that enables optimum control of the microstructure and hence mechanical properties. The inherent attributes of AM, however, have slowed the wide adoption of the technology in the industry. These include porosity, a non-ideal microstructure, a rough surface finish, and high residual stresses, which all affect the structural integrity of a part by significantly reducing the overall fatigue life[1].

This study aims to develop a deeper understanding of the fatigue crack growth rate of laser powder bed fusion (LPBF) produced Ti-6Al-4V from a microstructural perspective. In particular, the work focuses on fatigue life predictions by exploring the initiation, propagation and fast fracture and how these correlate to the material, process and build attributes. Considered are two experimental approaches: a stress-life approach measuring the cycles to failure for given applied cycling stress and a crack growth rate approach, measuring crack growth threshold values, crack growth propagation parameters and fast fracture (fracture toughness). Investigated are the influence of porosity, microstructure and residual stresses by performing various heat-treatments to allow for stress relaxation and microstructural transformations. Also, digital image correlation (DIC) is used to observe differences in crack growth behaviour between different material states.

Specimens were manufactured using two different LPBF machines. For both machines, Ti6Al4V-ELI powder was used with a lognormal particle size distribution of 35±15 µm in diameter. The five heat-treatment conditions are: as-build (AB), stress-relieved (SR) at 650 °C for 2 hours followed by a furnace cool, recrystallisation anneal (RA) at 920 °C for 2 hours followed by a furnace cool of 4 °C/min, hot isostatic press (HIP) at 920 °C for 2 hours at 1200 MPa followed by a furnace cool of 4 °C/min, and a duplex anneal (DA) at 910 °C for 2 hours followed by a water quench and an anneal at 800 °C followed by a furnace cool of 4 °C/min. Porosity was measured employing X-ray Computed Tomography using General Electric Phoenix V|Tome|X L240. Measurements are taken over the reduced section of approximately 5 × 5 mm, allowing for a voxel resolution of 5 µm^3. The microstructural analysis used a Zeiss MERLIN scanning electron microscope in conjunction with a backscatter detector.

As expected, higher porosity results in a significant reduction in fatigue life. A correlation exists between the pore distribution (Fig. 1a) and confidence bands; a higher distribution results in an increased scatter in stress-life fatigue data. Crack propagation follows the alpha lath orientation and tends to be transgranular (Fig. 1b). Distinct cracking planes are observed that align with the prior beta grain orientation. Due to the strong prior beta grain texture, the resultant mixed-mode type crack propagation is strongly orientation dependent. Also, measured is a residual crack opening mouth displacement (with the use of DIC). This is most apparent in AB samples with high residual stresses, however, was also observed after in the RA, HIP and DA conditions. Residual stress effects the initiation phase, and to a much lesser expend crack growth rates and final fracture. The statistical nature of porosity (and surface roughness) is such that a less favourably pores will initiate crack growth. Crack growth rate, however, is governed by extrinsic factors including the mixivity of the crack and plastic buildup behind the crack tip. This is mostly driven by crack close effects that are microstructurally dependent. Finally, the fineness of the microstructure governs fracture.

References:
Lui, S. and Shin, Y.C. (2018) Materials and Design. 164, 107552.

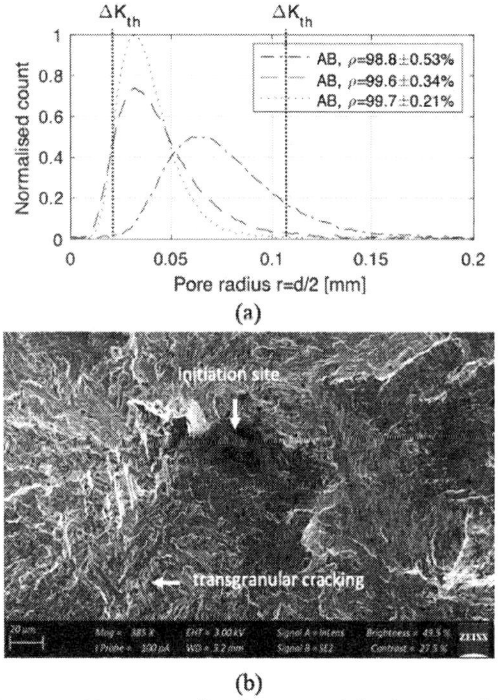

Figure 1. (a) Porosity distribution, and (b) fractography highlighting crack initiation and propagation mechanism.

Corresponding author: tbecker@sun.ac.za

NANOSWITCHES: TEM CHARACTERIZATION OF RESISTIVELY SWITCHING CHALCOGENIDES

J. Mayer[1,2], H. Du[1,2], A. Meledin[1,2] and V. Migunov[1,2]

[1]Ernst Ruska-Centre for Microscopy and Spectroscopy with Electrons, Forschungszentrum, Jülich, Germany, [2]Central Facility for Electron Microscopy, RWTH Aachen University, Aachen, Germany

The present work aims at a comprehensive understanding of the resistive switching mechanism in both oxide and phase change materials[1,2] on the basis of the local structural and bonding information at atomistic level obtained by aberration corrected (scanning) transmission electron microscopy ((S)TEM). For this purpose, we employ the advanced TEM and STEM facilities available at the Ernst Ruska-Centre for Microscopy and Spectroscopy with Electrons.

By employing aberration-corrected HRTEM and STEM, we conduct a detailed analysis of the atomistic structures of the different phases and defects, which are considered to play essential roles in the switching properties of oxides and other chalcogenides[3-5]. In our experiments, we aim at elucidating the atomistic structure of the filaments and elementary clusters which control the switching kinetics (Fig. 1). The results provide clear experimental evidence for the existence of multiple nanosized conducting filaments formed solely by atomic-scale point defects in a $SrTiO_3$ based thin-film ReRAM device. As visualized by means of STEM and EELS, V-shaped conducting filaments are formed with high concentrations of Ti^{3+} and O-vacancies. These two fundamental point defects lead to n-type conductivity and are found to be accompanied by significant amounts of Sr-deficiency and lattice distortions. Despite of the high concentration of point defect, the general perovskite lattice is structurally preserved, indicating that neither secondary phases nor extended defects are essential for the formation of conducting filaments.

Moreover, monochromated STEM enables us to obtain atomically resolved information on the electronic structures of the nanoscale defects and their coupling to the structural transitions[6]. Particularly, monochromated STEM enables us to obtain atomically resolved information on the electronic structure of the nanoscale defects and their coupling to the structural transitions. Moreover, we use the so-called fluctuation TEM technique to statistically analyse time-dependent nanodiffraction patterns. This technique makes it possible to investigate the short range and medium range order in amorphous materials and the unique crystallization kinetics of the phase change materials.

Furthermore, we use the so-called fluctuation TEM technique to statistically analyse time-dependent nanodiffraction patterns. This technique makes it possible to investigate the short range and medium range order in amorphous materials and the unique crystallization kinetics of the phase change materials. As a fundamental perspective, we aim at elucidating the formation and disruption of the atomistic structure of the filaments and elementary defects which essentially control the switching kinetics.

The authors acknowledge support by the German Research Foundation DFG in the framework of the Collaborative Research Centre SFB 917.

References:

1. R. Waser, R. Dittmann, M. Salinga, M. Wuttig (2010) Solid-State Electron. 54, 830.
2. R. Waser and M. Aono (2007) Nat. Mater. 6, 833.
3. M. Salinga *et al.* (2013) Nature Communications 4, 2371.
4. C. Baeumer *et al.* (2015) Nature Communications 6, 8610.
5. C. Rodenbücher *et al.* (2016) Scientific Reports 6, 32250.
6. H. Du *et al.* (2017) Chem. Mater. 29, 3164.

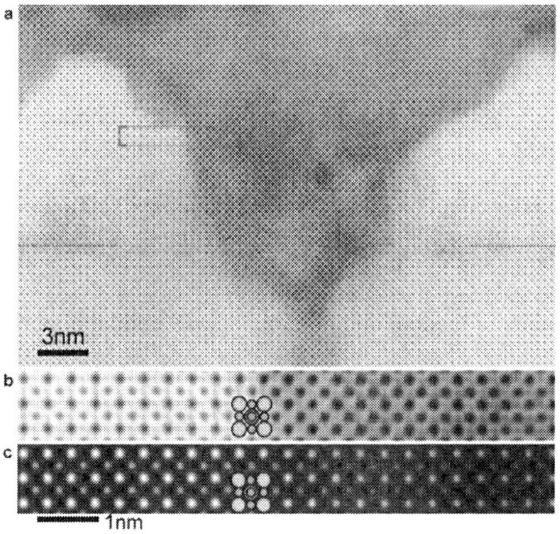

Figure 1. Atomic details of an area across the boundary between the film matrix and the filament. (a) High-resolution ABF image. The lines denote the interface between the film and the substrate. (b) Magnified part of the ABF image marked in (a). (c) The simultaneously acquired HAADF image corresponding to (b). Atomic columns are indicated by the projected unit cell of $SrTiO_3$ (Sr: green, Ti: red, O: blue).

Corresponding author: mayer@gfe.rwth-aachen.de

PRECIPITATE-MEDIATED DISLOCATION SOURCE MULTIPLICATION AND TRANSFORMATION

L.S. Feng[1], M.J. Mills[1] and Y. Wang[1]

[1]Department of Materials Science and Engineering, The Ohio State University, Columbus, USA

If we have a single Frank-Reed source in a crystal, we do not expect dislocations having Burgers vectors other than that of the F-R source to be generated. However, this may not be the case when the crystal is an ordered intermetallic phase. Ordered intermetallic phases are the essential constituents of an important class of structural materials, the high-temperature alloys, owing to their stability at elevated temperatures. Because of the low symmetry of their crystal structures as compared to those of their disordered counterparts, these ordered phases have rather complicated deformation mechanisms[1-4] that are difficult to uncover by experiment alone.

In this study we use a combination of ab initio calculation, phase field simulation and experimental characterization at individual dislocation level[3-4] to illustrate how the interactions between dislocations and precipitates of a low-symmetry ordered intermetallic phase such as γ'' ($D0_{22}$, tetragonal), the primary strengthening phase in IN718 superalloys and some HEAs, breed new dislocations having different Burgers vectors from the source dislocations. In particular, we show that the ordered precipitates serve as a "dislocation transformer" that generates new sources (new Burgers vectors) of dislocations from the original source. The new sources include Shockley partial dislocations and full dislocations, and the former could be responsible for the extensive micro-twinning observed in these alloys during deformation.

Figure 1 shows the interaction between the first ½ [0-11] (CA in Thompson notation) dislocation from a Frank-Reed source and the γ'' precipitate microstructure having three γ'' particles of different variants embedded in the γ matrix (dash ellipse in Frame 1, different colors represent three variants of the γ'' phase). Consecutive CA dislocations are generated from the bottom of the simulation box to mimic an active Frank-Reed source. For the first CA, it shears V2 and creates an APB-like fault (see Fig. 2 for the various stacking faults on the generalized stacking fault energy contour plot) within the precipitate (Frame 2). A δB partial is bred instantaneously to transform the unstable APB-like fault into an ISF within the precipitate (Frame 3)[3]. The δB partial is expanding into the matrix (Frame 3-6) under the applied load (shown in Frame 1). On the entrance side (down below the precipitates), a free δB partial is moving away from the precipitates (Frame 6), while on the exit side (up above the precipitate), a group of dislocations of CA+δB is moving away from the precipitates (Frame 6). The δB partial will be continuously bred as long as the Frank-Reed source at the bottom of the simulation box keeps operating. Multiple δB partials on the entrance side and multiple of CA, CB (Cδ+δB) and δB dislocations on the exit side will be generated. Note that δB and CB do not belong to the original Frank-Reed source (CA).

Understanding this precipitate-mediated dislocation source multiplication and transformation mechanism can further spark new ideas on alloy design, especially for tailoring the twinning mode. Some of these new ideas will be discussed. This work is supported by NSF under grant the DMREF program.

References:

1. Kovarik, L. *et al.* (2009) Prog. Mat. Sci. 54, 839.
2. McAllister, D. *et al.* (2016) Scripta Mater. 115. 108.
3. Lv, D.C., McAllister, D., Mills, M.J. and Wang, Y. (2016) Acta Mater. 118, 350.
4. Feng, L.S. *et al.* (2018) Acta Mater. 161, 99.

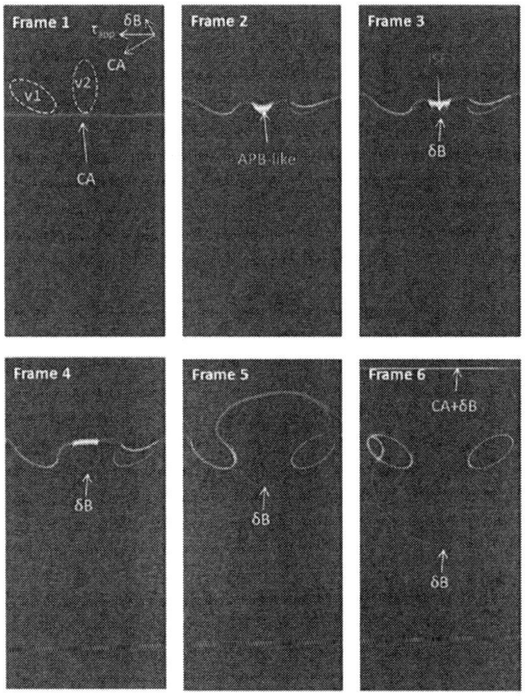

Figure 1. Interaction between the first CA dislocation and the precipitate microstructure.

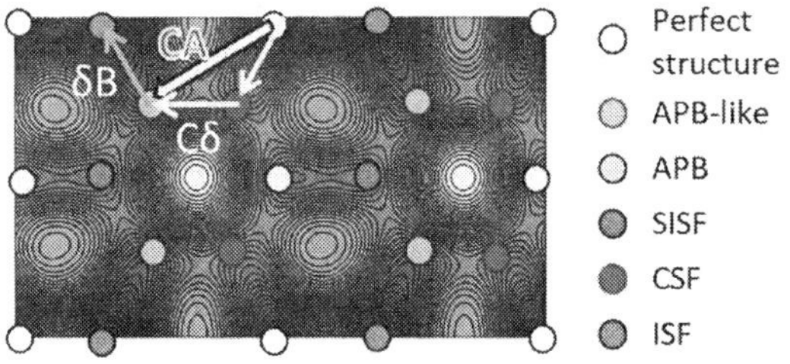

Figure 2. Generalized stacking fault energy surface of the g^2 phase. The arrows indicate the deformation pathway of the first CA.

Corresponding author: wang.363@osu.edu

PRIOR AUSTENITE TRANSFORMATION, IMAGE SEGMENTAION, CLUSTERING, AND OPTIMIZATION VIA GRAPH CUTS

S.R. Niezgoda[1,2], A.F. Brust[1] and E.J. Payton[3]

[1]Department of Materials Science and Engineering, The Ohio State University, Ohio, USA, [2]Department of Mechanical and Aerospace Engineering, The Ohio State University, Ohio, USA, [3]Air Force Research Laboratory, Ohio, USA

Problems involving image segmentation, atomic cluster identification, segmentation of microstructure constituents in images and austenite reconstruction have seen various approaches attempt to solve them with mixed results. No single computational technique has been able to effectively tackle these problems due to the vast differences between them. Prior grain reconstructions of the austenite to martensite transformation is hindered by the large volume of noise associated with the transformation. This can be attributed to the scale of the transformation, which results in a single prior austenite grain producing up to 24 martensitic variants; the plasticity associated with the massive formation of martensite; variations in the orientation relationship across variable compositions and morphologies; errors associated with the EBSD-indexing of the transformation microstructure; and annealing twins forming across the prior austenite microstructure. Due to the inherent noise associated with the transformation, reconstruction algorithms using point-to-point or flood-fill algorithms struggle to produce accurate and consistent reconstructions of the austenite microstructure.

We propose the application of graph cutting as a versatile technique that can provide solutions to numerous materials data analysis problems. This can be attributed to its configuration flexibility coupled with the ability to handle noisy experimental data. Implementation of a Bayesian statistical approach allows for the prior information, based on experimental results and already ingrained within nodes, to drive the expected solutions. This way, nodes within the graph can be grouped together with similar, neighboring nodes that are then assigned to a specific system with respect to calculated likelihoods. Associating probabilities with potential solutions and states of the system allows for quantitative, stochastic analysis. This technique can be applied to a number of inverse problems in materials science, such as image segmentation, microstructure phase and constituent segmentation, atomic cluster identification from atom probe tomography data sets, and the reconstruction of the parent microstructure from the EBSD-indexed post-transformation data set.

In order to fully flesh-out the algorithm for prior-austenite reconstruction, information related to the misorientation relationships between martensite variants associated with the same prior austenite grain or twin were utilized. Additionally, an accurate and automated measurement of the orientation relationship for the desired steel data sets was created through a Bayesian implementation and used conditionally within the reconstruction algorithm. Analysis of a number of steel and binary ferrous alloy data sets with variable orientation relationships was then performed to assess the range of the technique, along with segmentations on packet boundaries generated within a prior austenite grain[1]. It was found that the technique can sufficiently produce reasonable, probabilistically defined reconstructions of the prior austenite microstructure, even with the addition of copious amounts of noise. Additionally, it further separated itself from existing techniques through its ability to capture a number of prior austenite twins spanning across the parent grains. Validation was performed through both comparisons to an optical micrograph of the same chemically etched steel and comparison of retained austenite orientations associated with surrounding, reconstructed grains.

References:
1. Brust, A. *et al.* (2019) Metallurgical and Materials Transactions A 50(4), 837.

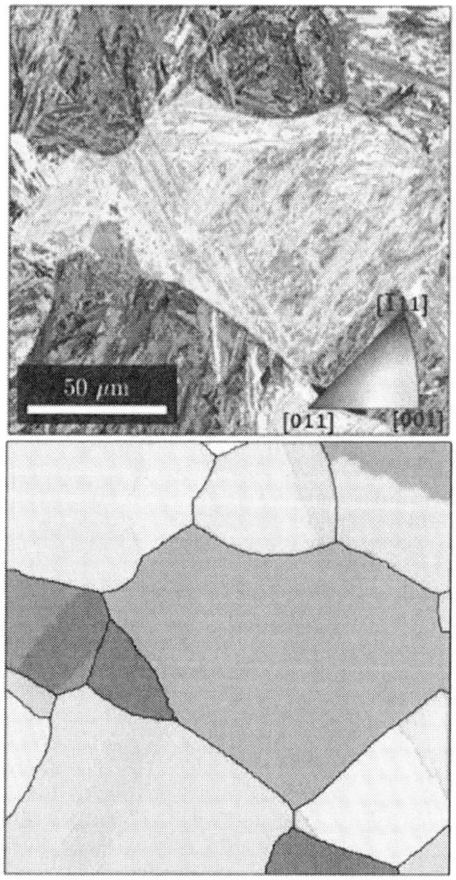

Figure 1. (Top) Post-transformation martensite microstructure for P122 12 wt% Cr steel sample. (Bottom) Automated reconstruction by graph cut algorithm.

Corresponding author: niezgoda.6@osu.edu

QUANTITATIVE STEM TOMOGRAPHY MEASUREMENTS OF INTERFACE ROUGHNESS FOR APERIODIC Mo-B4C MULTILAYERS SYSTEMS

G. Haberfehlner[1], G. Kothleitner[1], U. Heidorn[2], F. Hertlein[3], J. Wiesmann[3], T.H. Klassen[4] and W. Jaeger[5]

[1]Institute for Electron Microscopy and Nanoanalysis, Graz, Austria, [2]Graz Centre for Electron Microscopy, Graz, Austria, [3]Incoatec GmbH, Geesthacht, Germany, [4]Institute for Materials Research, Helmholtz-Zentrum, Geesthacht, Germany, [5]Institute for Materials Science, Kiel University, Kiel, Germany

The quantitative measurement of layer thickness and roughness of surfaces and interfaces at a sub-nanometer scale is instrumental for the functionality of device components based on layered material systems, such as X-ray optical systems or semiconductor devices. On a local scale, atomic force microscopy (AFM) is usually the method of choice for measuring roughness, of surfaces but AFM is unable to access buried interfaces. For buried interfaces, characterizations by transmission electron microscopy (TEM) are a useful tool, however, quantitative determination of interface roughness on a local scale is masked by the projection nature of TEM imaging. Electron tomography can overcome this limitation by reconstructing a 3D image of a sample from a tilt series of projections. Most frequently high-angle annular dark field (HAADF) imaging in scanning TEM (STEM) is applied for acquiring such data[1]. The challenges in quantitatively determining surface roughness of individual layers in multilayer systems by electron tomography are related to factors that can introduce uncertainties and distortions, such as tilt series alignment, choice of reconstruction algorithm and criterion for interface detection.

We have applied such methods to two different types of aperiodic Mo-B4C multilayer model systems with a nominal layer thickness varying between 1 and 2 nm. The 3D electron tomography employs a tilt series acquisition of STEM HAADF images taken from FIB-prepared cylinder-shaped samples, refined tilt alignment procedures, and novel algorithms for image reconstruction at high spatial 3D voxel resolution. The example of a two-dimensional (2D) surface reconstruction from a single interface (Fig. 1) depicts that layer interfaces can be accurately located and layer thickness distributions can be precisely measured. Thus, the roughness of individual interfaces of these ultrathin layers can be determined with variations lying well in the sub-nanometer range. Furthermore, mappings of 3D interface roughness profiles can be obtained and quantified for each layer individually and compared to 1D projection measurements.

Additional information about intermixing of elements between individual layers of the multilayer system as well as about processing-induced oxidation phenomena at the surface and at the substrate interface of the multilayers has been obtained from combining STEM HAADF imaging with annular bright field (ABF) imaging and with spectroscopic EELS and XEDS spectrum mappings of the projected elemental distributions. Thereby quantitative information about the complete structure of aperiodic multilayer systems for X-ray optics and of the distribution of interface roughness of the individual ultrathin layers can be extracted. Such quantitative data are instrumental for assessing and developing aperiodic multilayers for X-ray optics with optimized X-ray reflectivities.

Funding support by the European Union Seventh Framework Programme under Grant Agreement 312483 - ESTEEM2 (Integrated Infrastructure Initiative–I3) is gratefully acknowledged.

References:

1. Ercius, P. *et al.* (2015) Adv. Mater. 27, 5638.

2. Haberfehlner, G., Kothleitner, G., Heidorn, U., Hertlein, F., Wiesmann, J., Jäger, W. (2019) to be published.

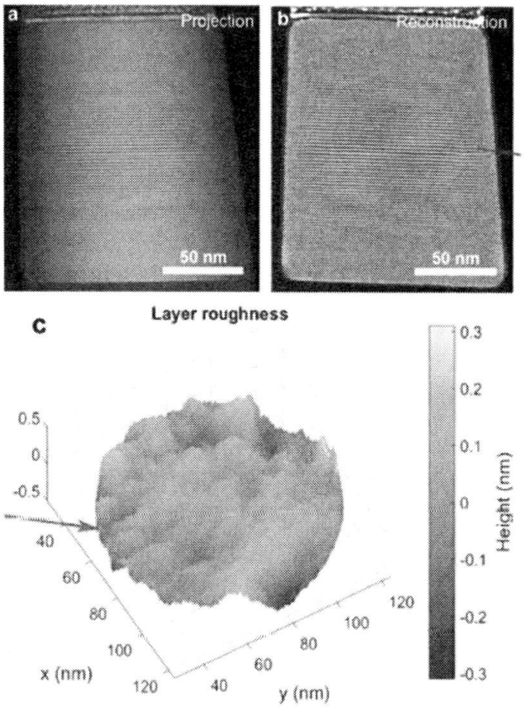

Figure 1. STEM tomography of an aperiodic multilayer system of alternating Mo and B4C layers. The cylindrically-shaped sample used for electron tomography has been prepared by a FIB-based method. (a) STEM HAADF projection, (b) Slice through reconstruction. (c) 2D surface reconstruction of a single interface.

Corresponding author: wolfgang.jaeger@tf.uni-kiel.de

STRUCTURE-TO-PROPERTY RELATIONSHIPS IN 9-12% Cr CSEF STEELS

J.E. Westraadt[1], G. Marx[1] and W.E. Goosen[1]

[1]Centre for HRTEM, Nelson Mandela University, Port Elizabeth

One of the major issues in characterising materials for industrial applications is the current need to *a priori* define the microstructural parameters that control the mechanical properties. Creep strength enhanced ferritic (CSEF) steels with 9-12% Cr are often used for critical high-temperature components in coal-fired power plants. The microstructure consists of hierarchically related internal interfaces[1], high dislocation densities, and secondary phases such as precipitates ($M_{23}C_6$, MX) that act as obstacles to dislocation movement[2] and stabilise the subgrain structure[3].

In CSEF steels the mean subgrain size (λ_{sg}), mean interparticle spacing (λ_p) and mean mobile dislocation density (ρ_m) are considered to be important for creep-strength[2]. In order to extract this information the chosen parameters are typically segmented from the experimental data collected from 2D sections and then transformed using stereological correction factors into 3D measurements. This mean-value approach neglects the spatial arrangement and size distributions of the different phases, which could be important for structure-to-property predictions. This study reports on the progress made towards a standardised quantitative characterisation procedure for CSEF steels and motivates the need for advanced statistical approaches such as n-point correlation function methods[4] to determine structure-to-property relationships in steels.

Table 1 shows a list of characterisation techniques that are used in combination to acquire the microstructural data from CSEF steels. A full description of the chacterisation procedures can be found in Marx[5]. This microstructural characterisation procedure was then applied to specimens extracted from different sections of a CSEF steel weldment[6]. Small punch creep (SPC) tests were conducted on 8 mm diameter and 0.5 mm thick disks at a load 276 N and 625 °C[6]. The parent plate (P91_PM) had a rupture time of 350 hours compared the heat affected zones (HAZ) samples that ruptured between 50-70 hours. Figure 1 shows the normalised strength contribution calculated from the microstructural parameters (λ_p, λ_{sg}, ρ_m) according to Holzer[2].

This result implies that mean precipitate spacing does not play an important role in the creep-strength since samples [X20_HAZ (1) and X20_HAZ (2)] with smaller λ_p ruptured within 70 hours. The strength calculations of Holzer[2], however, does not take the spatial arrangement of the precipitates relative to the grain boundaries into account. The HAZ samples had several precipitates located in the interior of the subgrains (Fig. 2), which could affect their ability to stabilise the subgrain structure. The use of n-point statistics to quantify the precipitate distribution relative to the grain boundaries and its use for structure-to-property relationships will be discussed.

Financial support from the Materials and Mechanics EPPEI Specialisation Group at UCT is gratefully acknowledged.

References:
1. Payton, E.J. *et al.* (2012) Scripta Materialia 66, 1045.
2. Holzer I. (2010) PhD Thesis, Graz University of Technology.
3. Kostka, A. *et al.* (2007) Acta Materialia 55(2), 539.
4. Simmons, J.P. *et al.* (2019) Statistical Methods for Material Science. Boca Raton, Florida, CRC Press.
5. Marx, G. (2019) PhD Thesis, Nelson Mandela University.
6. Westraadt, J.E. *et al.* (2019) Proc. Joint EPRI-123HiMAT, Nagasaki, 1109.

Table 1. Techniques used for the data acquisition.

Technique	Precipitates	"Grain-size"	Dislocations	Orientations	Time (min) (100 um[2])
EFTEM	$M_{23}C_6$, MX, Z-phase	no	no	no	20
STEM-EDS	$M_{23}C_6$, MX, Laves	yes	yes	no	6
TKD-EDS	All	yes	Limited	yes	46
BSE	$M_{23}C_6$, Laves, MX	yes	Limited	no	7
EBSD	Limited	yes	Limited	yes	4
XRD	Limited	yes	yes	Limited	N/A

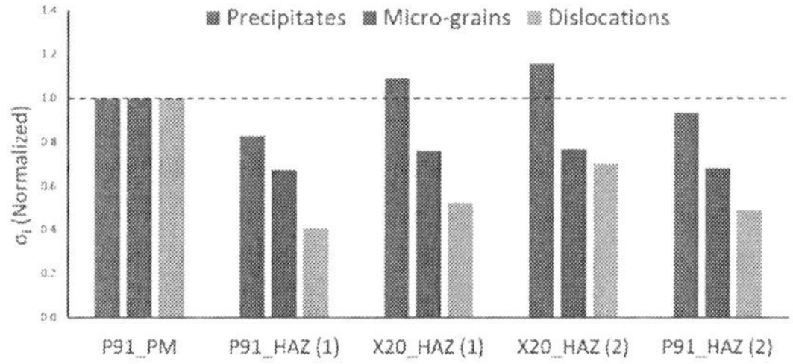

Figure 1. Normalised (relative to P91 parent material) microstructural parameter strength calculated according to Holzer[2].

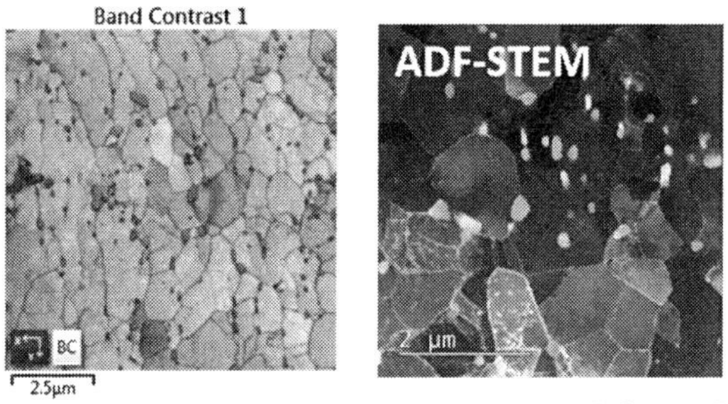

Figure 2. Band constrast TKD map (left) and ADF-STEM image (right) of the P91 heat affected zone sample.

Corresponding author: johan.westraadt@mandela.ac.za

STUDY OF RADIATION DAMAGE IN GaAs: FROM TEM TO Cs-CORRECTED HAADF STEM

J.H. Neethling[1], E.J. Olivier[1] and A Janse van Vuuren[1]

[1]Centre for HRTEM, Department of Physics, Nelson Mandela University, Port Elizabeth

Using Cs-corrected high angle annular dark field (HAADF) scanning transmission electron microscopy (STEM) for the study of radiation damage in GaAs is important since proton bombardment of GaAs, using doses in the range 10^{14} - 10^{15} H^+ cm^{-2}, has been used to create high-resistivity regions in GaAs devices[1]. GaAs is also a promising material for space solar cells, but the lattice defects induced by high-energy electron and proton irradiation in space cause a sharp decrease in power output of these solar cells[2]. Although the first author of this paper has investigated radiation damage and the formation and growth of dislocation loops in n-type GaAs since the 1980s, the earlier transmission electron microscopy (TEM) studies of radiation damage in proton bombarded GaAs[3,4], were carried out long before probe-Cs-corrected HAADF STEM imaging became available. This is a powerful technique for the determination of structural and chemical information at atomic resolution since the image characteristics do not depend on foil thickness and defocus value as for HRTEM. The n-type Si doped GaAs samples discussed in this paper were irradiated by fast neutrons and protons to total doses 2.8×10^{20} neutrons/cm^2 and 1×10^{17} protons/cm^2. HRSTEM specimens were prepared by using a Helios Nanolab 650 focused ion beam SEM and investigated in a double Cs-corrected JEOL ARM 200F operated at 200 kV.

Figure 1(a) shows a HAADF STEM image of a small {111} dislocation loop in neutron irradiated n-type GaAs. The plane of the interstitial loop consists of one layer of GaAs atoms while the stacking sequence across the loop is consistent with that of an extrinsic stacking fault. The insertion of a single GaAs rotated layer therefore results in the creation of two modified (rotated) layers. Figure 1(b) is a typical HAADF STEM lattice image of the {111} loop-like defect previously called a hydrogen platelet[5]. The single {111} GaAs layer of the interstitial loop forming the core of the hydrogen platelet is indicated by the arrow in (b) and the dark regions associated with the {111} plane of the defect viewed edge-on is indicative of lower density regions in GaAs, which is suggested to be due to small hydrogen filled bubbles along the {111} plane of the defect.

GaAs interstitials seem to agglomerate in a number of different configurations: interstitial edge dislocation loops on {110} planes[6], extrinsic Frank loops on {111} planes[6] and a type of hydrogen platelet proposed to be an interstitial loop on a {111} plane with hydrogen molecules in small bubbles adjacent to the {111} loop plane. In this study, no vacancy loops have been observed in irradiated and annealed n-type GaAs.

References:
1. Pearton, S.J. (1990) Mater. Sci. Rep. 4, 313.
2. Yamaguchi, M. (2001) Solar Energy Materials and Solar Cells 68, 31.
3. Neethling, J.H. and Snyman, H.C. (1986) J. Appl. Phys. 60, 941.
4. Neethling, J.H., Snyman, H.C. and Ball, C.A.B. (1988) J. Appl. Phys. 63, 704.
5. Neethling, J.H. *et al.* (2019) Acta Physica Polonica A (*in press*).

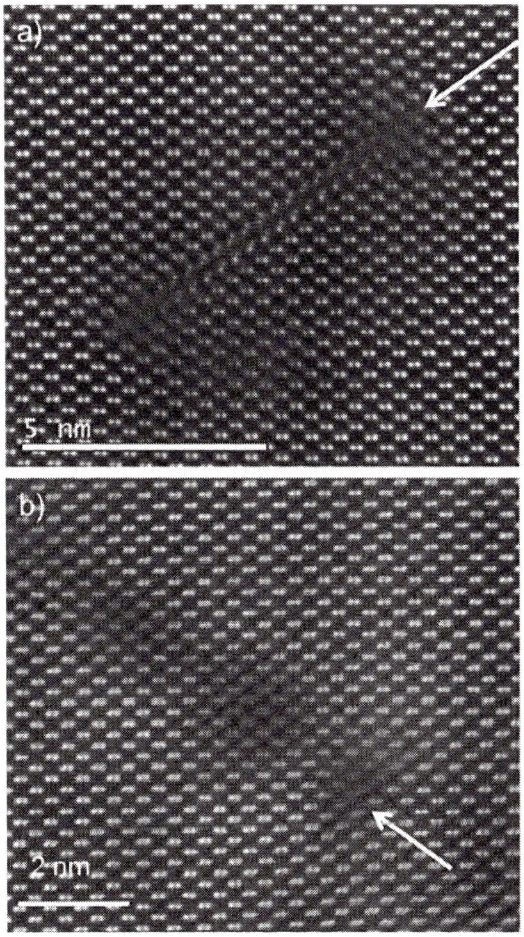

Figure 1. (a) HAADF STEM image of a dislocation loop on a {111} plane in neutron irradiated n-type GaAs. The plane of the interstitial loop (indicated by the arrow) consists of one layer of GaAs atoms while the stacking sequence across the loop is consistent with that of an extrinsic stacking fault. (b) Typical HAADF STEM lattice image of the loop-like defect, previously called a hydrogen platelet, associated with hydrogen molecules. The single {111} GaAs layer of the interstitial loop forming the core of the hydrogen platelet is indicated by the arrow in (b). The electron beam direction is <110>.

Corresponding author: Jan.Neethling@mandela.ac.za

THE ROLE OF STRUCTURAL AND COMPOSITIONAL INSTABILITIES ON THE MICROSTRUCTURAL EVOLUTION OF METASTABLE BETA TITANIUM ALLOYS

Y. Zheng[1], R. Banerjee[2], D. Banerjee[3], G.B. Viswanathan[1], Y. Wang[1] and H.L. Fraser[1]

[1]CAMM, The Ohio State University, Columbus, USA, [2]Materials Science and Engineering, University of North Texas, Denton, USA, [3]Materials Engineering, Indian Institute of Science, Bangalore, India

In the current work, in order to improve the overall strength, precipitation strengthening by *hcp* structure alpha phase precipitation in *bcc* structure beta phase matrix is effected in a metastable beta titanium alloy, Ti-5Al-5Mo-5V-3Cr (Ti-5553), exploiting differing roles of the structural and compositional instabilities (associated with the athermal and isothermal omega phase of *hexagonal* structure) during specifically designed heat treatments[1-3]. The microstructural evolution has been explored using advanced electron microscopy, and validated using phase field simulation[1-3].

The first set of experiments involved samples of Ti-5553 (having first been solution heat treated and quenched) being heated to 600 °C at the heating rate of 100 °C/min and isothermally held for 2 hours before being fast cooled to room temperature. This heat treatment produced a fine distribution of the alpha phase and aerial number density of alpha phase precipitates was analyzed using MIPAR™ software package[4] as ~2 ppts/um^2. The refined alpha microstructure appears to be very similar to those formed by step-quenching from above the beta transus to 600°C[5], formed via the pseudo-spinodal mechanism[6].

The second set of experiment involved samples being heated to 600 °C at a much slower rate, 5 °C/min, and isothermally held for 2 hours before being fast cooled to room temperature. A super-refined alpha microstructure was produced with a much higher aerial number density[2] of 40 ppts/um^2. This super-refined alpha microstructure is formed by the direct influence of pre-formed isothermal omega phase particles shown in the HAADF-STEM image in Fig. 1(a). The pre-formed isothermal omega phase particles can change the local concentration (by rejecting solutes of Al, Mo, V and Cr into beta matrix) and structure (from *bcc* to *hexagonal*), shown in Fig. 1(b-c)[2]. Therefore, isothermal omega phase can provide extra driving force for subsequent alpha phase precipitation and pre-formed omega/beta interface can act as favorable nucleation sites for the alpha phase. The large number of isothermal omega phase particles provide the large number sites for the nucleation of super-refined alpha precipitates.

The third set of experiments involved samples heated to 350 °C at a rate of 5 °C/min, and subsequently up-quenched to 600 °C and isothermally held for 2 hours before being fast cooled to room temperature. The produced microstructure exhibits an aerial number density in between as ~20 ppts/um^2, which is named as more-refined alpha microstructure[3]. Such more-refined alpha microstructure was produced via the indirect influence of pre-formed isothermal omega phase particles. Isothermal omega phase was formed firstly during isothermal aging at 350 °C and fine scale alpha phase precipitates nucleated and grew after long time aging at 350 °C. But the fine scale alpha phase precipitates transformed back to beta phase matrix during up-quenching to 600 °C and also created solute inhomogeneity in beta matrix that would restrain the nucleation of subsequent alpha precipitation at 600 °C[3]. Therefore, the number density of more-refined alpha microstructure was directly influenced by the compositional inhomogeneity in the beta phase matrix which was indirectly influenced by the pre-formed isotheral omega phase[3].

References:
1. Zheng, Y. *et al.* (2016) Scripta Materialia 111, 81.
2. Zheng, Y. *et al.* (2016) Acta Materialia 103, 850.
3. Zheng, Y. *et al.* (2016) Acta Materialia 103, 165.
4. Sosa, J.M., Huber, D.E., Welk, B. and Fraser, H.L. (2014) Integrated Materials and Manufacturing Innovation 3, 10.
5. Nag, S. (2012) Acta Materialia 60, 6247.
6. Boyne, A. *et al.* (2014) Acta Materialia 64, 188.

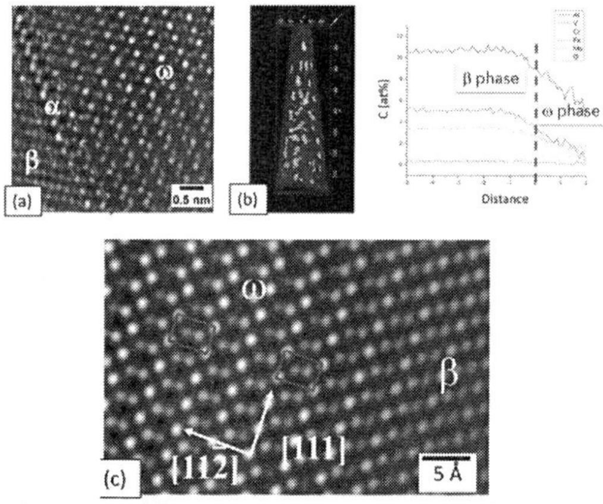

Figure 1. (a) HAADF-STEM image showing the pre-formed omega/beta interface acting as nucleation site of super-refined alpha precipitate; (b) Atom probe reconstruction showing isothermal omega particle and composition profile of Al, V, Mo, Cr, O, Fe across the omega/beta interface; (c) HAADF-STEM image showing the structure of isothermal omega particle[3].

Corresponding author: fraser.3@osu.edu

THREE DIMENSIONAL SPATIALLY RESOLVED ACOUSTIC SPECTROSCOPY (3D-SRAS)

T.K. Ales[1], M. Clark[2] and P.C. Collins[1]

[1]Department of Materials Science and Engineering, Iowa State University, Ames, Iowa, USA, [2]Dept. of Applied Optics, University of Nottingham, Nottingham, UK

Spatially resolved acoustic spectroscopy (SRAS) is emerging as a new way to characterize the microstructure of a material, but over a large length scale. The method was pioneered at the University of Nottingham (UK) by researchers in the Optics and Photonics research group. It involves the alignment of two lasers, one of which is used to induce a transient material response in the material and the second which measures characteristics of the transient response.

Specifically, in a prototypical configuration, a Q-switched 1064 nm infrared laser is passed through an optical grating that is imaged onto the sample surface. This pulsed laser impulse creates periodic lines of energy maxima, generating multiple modes of acoustic wave propagation, the most useful of which is a surface acoustic wave (SAW). This SAW then moves along the specimens surface, elastically displacing the crystalline material a few nanometers, at a velocity of approximately 10,000 kilometers per hour. The precise velocity is directly related to the nature of the elastic stiffness tensor (cij), which means that if multiple measurements of the velocity are conducted, the orientation of the elastic stiffness tensor relative to the surface can be determined, meaning that the crystal orientation can be determined. This velocity of the SAW is measured by the second laser, typically a green (532 nm) continuous laser that is focused onto the surface, and is perturbed by the SAW. This perturbation is captured as the green detection laser "wobbles" across a detector consisting of multiple photodiodes configured in an electrically balanced manner. Coupling the lasers and optics with an oscilloscope and stage scanning system, it is possible to obtain orientation microscopy information without the need for an electron gun. Additional details regarding this method can be found in the literature[1-3].

The advantages of this exciting new technique are that very large area orientation microscopy datasets can be obtained in a relatively short amount of time. Figure 1, for example, is an orientation map of a slice through a large-scale additively manufactured Ti-6Al-4V test article. The area scanned, approximately 14,500 mm^2, was characterized in less than 24 hours.

While this technique is exciting, until recently, it had been limited to one research lab in the University of Nottingham, owing, in part, to the complexities during the commissioning of such systems. Thus, many of the possibilities remain largely unexplored. We have recently commissioned the first system outside of the UK, and are adapting the technology to integrate into a serial sectioning instrument, affording the possibilities of three-dimensional large-scale spatially resolved orientation information using SAWs. Because the method provides measures of physical properties, from which orientation is determined, it also has the potential to provide additional information such as the composition, which will influence the elastic stiffness tensor. While we are pursuing 3D information, we are also working on improving the spatial resolution of this technique, and are tracking related work which uses a similar approach sans the stage scanning to determine time-resolved changes in physical properties for materials subjected to external stimuli.

We will present on this new technique and some of the exciting new experiments that may be conducted on such systems.

References:
1. Smith, R. J. *et al.* (2014) Measurement Science and Technology 25(5), 055902.
2. Li, W. *et al.* (2012) The Journal of the Acoustical Society of America 132(2), 738.
3. Sharples, S. D., Clark, M. and Somekh, M. G. (2006) Optics express 14(22), 10435.

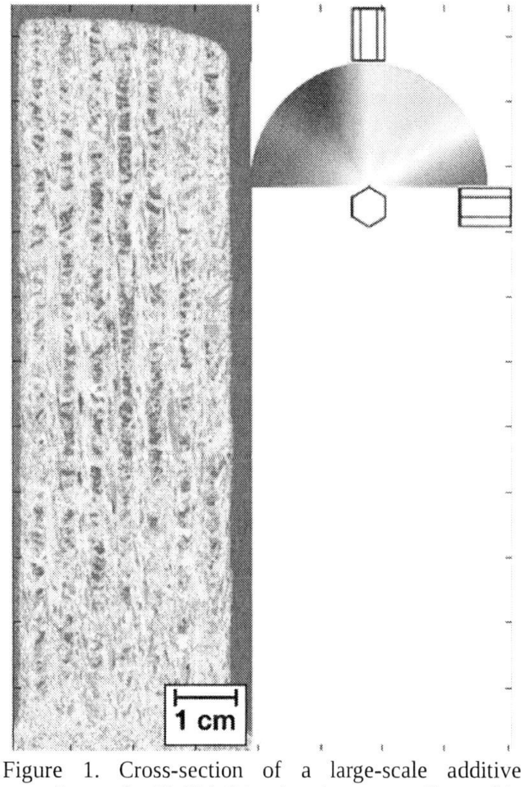

Figure 1. Cross-section of a large-scale additive manufactured Ti-6Al-4V showing crystallographic orientation.

Corresponding author: pcollins@iastate.edu

TRANSFORMATIONS, RECRYSTALLIZATION, MICROTEXTURE AND PLASTICITY IN TITANIUM ALLOYS: AN EMERGING VIEW

D. Banerjee

Advanced Facility for Microscopy and Microanalysis and Materials Engineering, Indian Institute of Science, Bangalore, India

We first describe different facets of the β to α transformation in titanium alloys. α morphology and distribution have been mapped as a function of β stabilizer content and aging temperature[1]. A dominant transgranular grouping of α variants with a common close packed direction characterizes transformation patterns across the entire range of β stabilizer content. Variant distribution has been quantified by EBSD techniques and explained through analysis of stress accommodation by different variant groups[2].

Recrystallization during thermomechanical processing in the α/β regime has been studied with the help of fine scale, orientation image mapping techniques using precession electron diffraction. We find two distinct recrystallization processes. The first is associated with the well-known α globularisation process. An alternative, dynamic recrystallization process is also observed in which newly recrystallized α and β grains form with the Burgers orientation relationship with each other. We call this process epitaxial recrystallization[3].

We then present the role of transformation and recrystallisation texture on the plasticity of titanium alloys through a quantitative EBSD /TEM based analysis of slip systems in the α phase and slip transfer between the α and β phases[4].

We first examine slip in equiaxed α and transformed β constituent of the duplex structure in the high temperature alloys Ti6242 and Ti6246. For slip analysis, the compressed samples were examined by SEM and EBSD. Using the secondary electron (SE) and backscattered electron (BSE) micrographs of the slip lines and EBSD patterns from the scanned regions, permitted slip system analysis from the surface offsets associated with slip bands on the electropolished surfaces. This data has been used to obtain the ratios of CRSS on different slip systems by a methodology described by Li and co-workers[5].

We also examine the effect of slip transfer from the softer, equiaxed alpha phase in determining the operative slip systems in transformed β, in the context of the local texture as determined by the dynamic recrystallisation process describe earlier.

Finally, we examine plasticity in microstructures with fine scale, multivariant distributions of α in a β matrix (Fig. 1), as is typical of the high strength alloy Ti5553. We show how the variant distribution determines the slip length and defines the relationship between slip length and room temperature yield strength of these alloys[6]. Strain partitioning between α and β is explored through high resolution DIC techniques.

References:
1. Choudhuri, D. *et al.* (2015) Proceedings of the 13th World Conference on Titanium, Ed. V Venkatesh *et al.* TMS (The Minerals, Metals & Materials Society), 515.
2. Balachandran, S. *et al.* (2016) Acta Mater. 106, 374.
3. Balachandran, Kumar, K. and Banerjee, D. (2017) Acta Mater. 13, 1423.
4. Banerjee, A. *et al.* (2015) Proceedings of the 13th World Conference on Titanium, Ed. V Venkatesh *et. al*, TMS (The Minerals, Metals & Materials Society), 1095.
5. Li, H. *et al.* (2013) Acta Mater. 61, 7555.
6. Chinara, M. (2018) Masters Thesis, Indian Institute of Science, Bangalore, India.

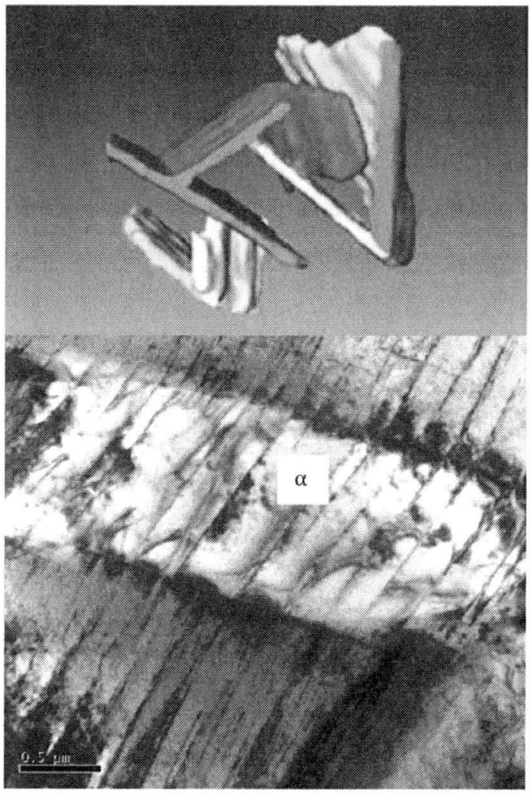

Figure 1. Multivariant α structures and slip transfer between the β and α phases

Corresponding author: dbanerjee@iisc.ac.in

UNAMBIGUOUS EBSD INDEXING OF COMPLEX MULTI-PHASE MICROSTRUCTURES

M. De Graef

Department of Materials Science and Engineering, Carnegie Mellon University, Pittsburgh, USA

Many modern engineering materials have microstructures with features that extend all the way down to the submicron length scale. In backscattered electron diffraction (EBSD) experiments in a scanning electron microscope (SEM), multiple phases may be intermixed at a length scale smaller than the size of the beam interaction volume; hence, the observed diffraction patterns will be an overlap pattern with contributions from all the phases present in the interaction volume. Such patterns pose problems for the commercial indexing packages based on the Hough transform; this transforms extracts the positions and orientations of individual Kikuchi bands from the pattern, but does not have any way to determine which phase each band belongs to.

Recent development of two whole-pattern indexing algorithms now makes it possible to disentangle the overlapping diffraction patterns, and to extract quantitative information about all the phases present in the interaction volume. In whole-pattern indexing, a forward model is used to predict the intensity distribution in EBSD patterns; one can then use pattern matching approaches to identify where the experimental pattern is located on the Kikuchi sphere, thereby obtaining the orientation of the crystal volume that gave rise to the pattern. Dictionary Indexing[1] and Spherical Indexing[2,3] are two recent techniques that can be used for whole-pattern indexing; we will briefly describe both techniques and discuss their most important applications and differences. We will apply both techniques to the case of fine α-β microstructures in a Ti-10-2-3 alloy.

Both DI and SI algorithms use a so-called EBSD master pattern, which represents the backscattered electron yield on a spherical surface surrounding a hypothetical spherical crystal with an electron source at its center. Since the orientation relation (OR) between the two phases is well know (Burgers OR), one can generate an artificial EBSD master pattern by rotating one pattern into the orientation described by the OR, thus obtaining an overlap master pattern that can be used for indexing. Fig. 1 shows the overlap master pattern for a mixture of 35 vol% β phase and the remainder α phase; the α phase master pattern was rotated by the following OR: $[111]_\beta$ // $[11.0]_\alpha$ and $(1\ \text{-}10)_\beta$ // $(00.1)_\alpha$. This master pattern can be used to directly index EBSD patterns; since the cubic phase master pattern is oriented in the standard setting, the Euler angles resulting from the indexing reflect the orientation of the cubic phase. In basket weave or colony regions, overlap patterns are typically observed and they can be predicted by means of the overlap master pattern in Fig. 1. When the indexing is carried out using triclinic symmetry (i.e., no symmetry), then the orientation of the β phase can be reduced to the fundamental zone (FZ) for the cubic phase, resulting directly in the orientation of the parent β grain. We will show that when the orientation obtained from the triclinic indexing step is concatenated with the inverse of the cubic orientation in the FZ, one obtains one of the cubic symmetry operators. Comparison of this operator with a pre-computed list then immediately identifies the α phase variant that is present in the interaction volume. Thus, triclinic indexing with respect to the overlap master pattern simultaneously produces the orientation of the parent β phase and uniquely identifies each of the 12 α phase variants (Fig. 2). We will discuss the algorithm that accomplishes this result as well as its implications for other microstructure types for which overlap EBSD patterns are commonly observed, for instance finely twinned microstructures or very fine phase mixtures in steels.

References:

1. Chen, Y.N. *et al.* (2015) Microsc. MicroAnal. 21, 739.
2. Lenthe, W.C., Singh, S., and De Graef, M. (2019) Ultramicroscopy (in press)
3. Lenthe, W.C. *et al.* (2019) Acta Materialia (under review)

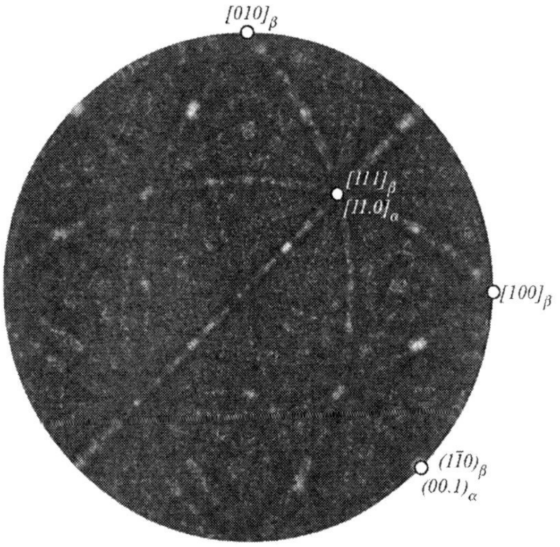

Figure 1. Overlap master pattern (at 8 kV) for Ti $\alpha-\beta$ phases in the Burgers OR.

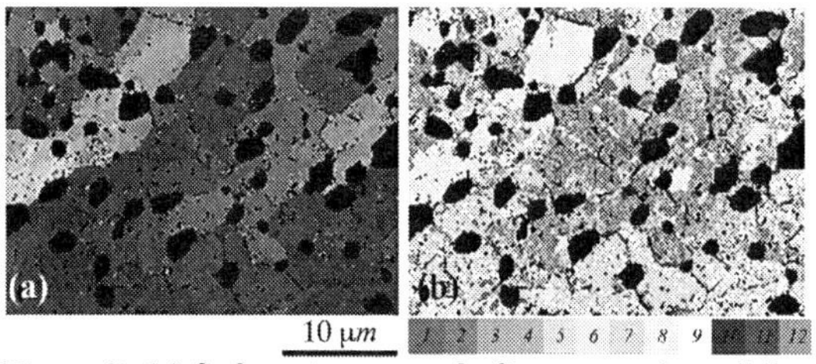

Figure 2. (a) β phase inverse pole figure map (α nodules masked out); (b) color map of the 12 α phase orientation variants present in the basket weave regions.

Corresponding author: degraef@cmu.edu

Author Index